tredition

tredition was established in 2006 by Sandra Latusseck and Soenke Schulz. Based in Hamburg, Germany, tredition offers publishing solutions to authors and publishing houses, combined with worldwide distribution of printed and digital book content. tredition is uniquely positioned to enable authors and publishing houses to create books on their own terms and without conventional manufacturing risks.

For more information please visit: www.tredition.com

TREDITION CLASSICS

This book is part of the TREDITION CLASSICS series. The creators of this series are united by passion for literature and driven by the intention of making all public domain books available in printed format again - worldwide. Most TREDITION CLASSICS titles have been out of print and off the bookstore shelves for decades. At tredition we believe that a great book never goes out of style and that its value is eternal. Several mostly non-profit literature projects provide content to tredition. To support their good work, tredition donates a portion of the proceeds from each sold copy. As a reader of a TREDITION CLASSICS book, you support our mission to save many of the amazing works of world literature from oblivion. See all available books at www.tredition.com.

 Project Gutenberg

The content for this book has been graciously provided by Project Gutenberg. Project Gutenberg is a non-profit organization founded by Michael Hart in 1971 at the University of Illinois. The mission of Project Gutenberg is simple: To encourage the creation and distribution of eBooks. Project Gutenberg is the first and largest collection of public domain eBooks.

Priestley in America 1794-1804

Edgar Fahs Smith

Imprint

This book is part of TREDITION CLASSICS

Author: Edgar Fahs Smith
Cover design: Buchgut, Berlin – Germany

Publisher: tredition GmbH, Hamburg - Germany
ISBN: 978-3-8472-2838-7

www.tredition.com
www.tredition.de

Copyright:
The content of this book is sourced from the public domain.

The intention of the TREDITION CLASSICS series is to make world literature in the public domain available in printed format. Literary enthusiasts and organizations, such as Project Gutenberg, world-wide have scanned and digitally edited the original texts. tredition has subsequently formatted and redesigned the content into a modern reading layout. Therefore, we cannot guarantee the exact reproduction of the original format of a particular historic edition. Please also note that no modifications have been made to the spelling, therefore it may differ from the orthography used today.

PREFACE

The writer, in studying the lives of early American chemists, encountered the name of *Joseph Priestley* so frequently, that he concluded to institute a search with the view of learning as much as possible of the life and activities, during his exile in this country, of the man whom chemists everywhere deeply revere. Recourse, therefore, was had to contemporary newspapers, documents and books, and the resulting material woven into the sketch given in the appended pages. If nothing more, it may be, perhaps, a connecting chapter for any future history of chemistry in America. Its preparation has been a genuine pleasure, which, it is hoped by him whose hand guided the pen, will be shared by his fellow chemists, and all who are interested in the growth and development of science in this country.

PRIESTLEY IN AMERICA

There lies before the writer a tube of glass, eleven and one half [Pg 1] inches in length and a quarter of an inch in diameter. Its walls are thin. At one end there is evidence that an effort was made to bend this tube in the flame. Ordinarily it would be tossed aside; but this particular tube was given the writer years ago by a great-grandson of Joseph Priestley. Attached to the tube is a bit of paper upon which appear the words "piece of tubing used by Priestley." That legend has made the tube precious in the heart and to the eye of the writer. Everything relating to this wonderful figure in science, history, religion, politics and philosophy is very dear to him. On all sides of him are relics and reminders of Priestley. Not all, but many of his publications are near at hand. After perusal of these at various times, and while reading the many life sketches of Priestley, there has come the desire to know more [Pg 2] about his activities during the decade (1794-1804) he lived in America. Isn't it fair to declare that the great majority of chemical students think of Priestley as working only in England, his native land, and never give thought to his efforts during the last ten years of his life? It has been said that he probably inspired and incited the young chemists of this country to renewed endeavor in their science upon his advent here. There is no question that he influenced James Woodhouse and his particular confreres most profoundly, as he did a younger generation, represented by Robert Hare. Priestley again set in rapid motion chemical research in the young Republic. [1] He must therefore have done something himself. What was it? Is it worth while to learn the character of this work? Modern tendencies are antagonistic to the past. Many persons care nothing for history. It is a closed book. They do not wish it to be opened, and yet the present is built upon the early work. In reviewing the development of chemistry in this country everything, from the first happening here, should be laid upon the table for study and reflection. Thus believing, it will not be out of place to seek some light upon the occupation of the discoverer of oxygen after he came to live among us — with our fathers. [Pg 3]

Noble-hearted, sympathetic Thomas E. Thorpe wrote:

If, too, as you draw up to the fire 'betwixt the gloaming and the mirk' of these dull, cold November days, and note the little blue flame playing round the red-hot coals, think kindly of Priestley, for he first told us of the nature of that flame when in the exile to which our forefathers drove him.

Right there, "the nature of the flame," is one thing Priestley did explain in America. He discovered carbon monoxide—not in England, but in "exile." [2] It may not be an epoch-making observation. There are not many such and those who make them are not legion in number. It was an interesting fact, with a very definite value, which has persisted through many succeeding decades and is so matter-of-fact that rarely does one arise to ask who first discovered this simple oxide of carbon.

Priestley was a man of strong human sympathies. He loved to mingle with men and exchange thoughts. Furthermore, Priestley was a minister—a preacher. He was ordained while at Warrington, and gloried in the fact that he was a [Pg 4] Dissenting Minister. It was not his devotion to science which sent him "into exile." His advanced thought along political and religious lines, his unequivocal utterances on such subjects,—proved to be the rock upon which he shipwrecked. It has been said—

By some strange irony of fate this man, who was by nature one of the most peaceable and peace-loving of men, singularly calm and dispassionate, not prone to disputation or given to wrangling, acquired the reputation of being perhaps the most cantankerous man of his time....

There is a wide-spread impression that Priestley was a chemist. This is the answer which invariably comes from the lips of students upon being interrogated concerning him. The truth is that Priestley's attention was only turned to chemistry when in the thirties by Matthew Turner, who lectured on this subject in the Warrington Academy in which Priestley labored as a teacher. So he was rather advanced in life before the science he enriched was revealed to him in the experimental way. Let it again be declared, he was a teacher. His thoughts were mostly those of a teacher. Education occupied him. He wrote upon it. [Pg 5] The old Warrington Academy was a "hot-bed of liberal dissent," and there were few subjects upon which he did not publicly declare himself as a dissenter.

He learned to know our own delightful Franklin in one of his visits to London. Franklin was then sixty years of age, while Priestley was little more than half his age. A warm friendship immediately sprang up. It reacted powerfully upon Priestley's work as "a political thinker and as a natural philosopher." In short, Franklin "made Priestley into a man of science." This intimacy between these remarkable men should not escape American students. Recall that positively fascinating letter (1788) from Franklin to Benjamin Vaughan, in which occur these words:

Remember me affectionately ... to the honest heretic Dr. Priestley. I do not call him honest by way of distinction, for I think all the heretics I have known have been virtuous men. They have the virtue of Fortitude, or they would not venture to own their heresy; and they cannot afford to be deficient in any of the other virtues, as that would give advantage to their many enemies.... Do not however mistake me. It is not to my good [Pg 6] friend's heresy that I impute his honesty. On the contrary 'tis his honesty that has brought upon him the character of heretic.

Much of Priestley's thought was given to religious matters. In Leeds he acknowledged himself a *humanitarian*, or

a believer in the doctrine that Jesus Christ was in nature solely and truly a man, however highly exalted by God.

His home in Leeds adjoined a "public brew house." He there amused himself with experiments on carbon dioxide (fixed air). Step by step he became strongly attracted to experimentation. His means, however, forbade the purchase of apparatus and he was obliged to devise the same and also to think out his own methods of attack. Naturally, his apparatus was simple. He loved to repeat experiments, thus insuring their accuracy.

In 1772 he published his first paper on Pneumatic Chemistry. It told of the impregnation of water with carbon dioxide. It attracted attention and was translated into French. This soda-water paper won for Priestley the Copley medal (1773). While thus signally honored he continued publish [Pg 7] ing views on theology and metaphysics. These made a considerable uproar.

Then came the memorable year of 1774—the birth-year of oxygen. How many chemists, with but two years in the science, have been so fortunate as to discover an element, better still probably the most important of all the elements! It was certainly a rare good fortune! It couldn't help but make him the observed among observers. This may have occasioned the hue and cry against his polemical essays on government and church to become more frequent and in some instances almost furious.

It was now that he repaired to London. Here he had daily intercourse with Franklin, whose encouragement prompted him to go bravely forward in his adopted course.

It was in 1780 that he took up his residence in Birmingham. This was done at the instance of his brother-in-law. The atmosphere was most congenial and friendly. Then, he was most desirous of resuming his ministerial duties; further, he would have near at hand good workmen to aid him in the preparation of apparatus for his philosophical pursuits. Best of all his friends were there, including those devoted to science. Faujar St. Fond, a French geologist has recorded a visit to Priestley [Pg 8] —

Dr. Priestley received me with the greatest kindness.... The building in which Dr. Priestley made his chemical and philosophical experiments was detached from his house to avoid the danger of fire. It consisted of several apartments on the ground floor. Upon entering it we were struck with a simple and ingenious apparatus for making experiments on inflammable gas extracted from iron and water reduced to vapour.

If, only, all the time of Dr. Priestley in Birmingham had been devoted to science, but alas, his "beloved theology" claimed much of it. He would enter into controversy—he would dissent, and the awful hour was advancing by leaps and bounds. The storm was approaching.

It burst forth with fury in 1791. The houses of worship, in which he was wont to officiate, were the first to meet destruction, then followed his own house in which were assembled his literary treasures and the apparatus he had constructed and gathered with pains, sacrifice and extreme effort. Its demolition filled his very soul with deepest sorrow. Close at hand, the writer has a neat little chemical balance. It was brought to this country by Priestley, and tradition has it, [Pg 9] that it was among the pieces of the celebrated collection of chemical utensils rescued from the hands of the infuriated mob which sought even the life of Priestley, who fortunately had been spirited or hidden away by loyal, devoted friends and admirers. In time he ventured forth into the open and journeyed to London, and when quiet was completely restored, he returned to one of his early fields of activity, but wisdom and the calm judgment of friends decided this as unwise. Through it all Priestley was quiet and philosophical, which is evident from the following story:

A friend called on him soon after the riots and condoled with him for his loss in general, then mentioned the destruction of his books as an object of particular regret. Priestley answered, "I should have read my books to little purpose if they had not taught me to bear the loss of them with composure and resignation."

But the iron had entered his soul. He could not believe that in his own England any man would be treated as he had been treated. His country was dear to him. He prized it beyond expression, but he could not hope for the peace his heart craved. His family circle was broken, two of his sons hav [Pg 10] ing come to America, so in the end, deeply concerned for his life-companion's comfort, the decision to emigrate was reached, and their faces were turned to the West.

In reviewing the history of chemistry the remark is frequently heard that one blotch on the fair escutcheon of French science was placed there when the remorseless guillotine ushered Lavoisier into eternity. Was not the British escutcheon of science dimmed when Priestley passed into exile? Priestley—who had wrought so splendidly! And yet we should not be too severe, for an illustrious name—Count Rumford—which should have been ours—was lost to us by influences not wholly unlike those which gained us Priestley. Benjamin Thompson, early in life abandoned a home and a country which his fellow citizens had made intolerable.

Read Priestley's volumes on Air and on Natural Philosophy. They are classics. All conversant with their contents agree that the experimental work was marvelous. Priestley's discovery of oxygen was epoch-making, but does not represent all that he did. Twice he just escaped the discovery of nitrogen. One wonders how this occurred. He had it in hand. The other numerous observations made by him antedate his American life and need not be mentioned [Pg 11] here. They alone would have given him a permanent and honorable rank in the history of chemistry. Students of the science should reserve judgment of Priestley until they have familiarized themselves with all his contributions, still accessible in early periodicals. When that has been done, the loss to English science, by Priestley's departure to another clime will be apparent.

His dearest friends would have held him with them. Not every man's hand was against him—on the contrary, numerous were those, even among the opponents of his political and theological utterances, who hoped that he would not desert them. They regretted that he had—

turned his attention too much from the luminous field of philosophic disquisition to the sterile regions of polemic divinity, and the still more thorny paths of polemic politics....

from which the hope was cherished that he would recede and devote all his might to philosophical pursuits.

A very considerable number ... of enlightened inhabitants, convinced of his [Pg 12] integrity as a man, sincerity as a preacher, and superlative merit as a philosopher, were his strenuous advocates and admirers.

But the die had been cast, and to America he sailed on April 8, 1794, in the good ship *Sansom*, Capt. Smith, with a hundred others—his fellow passengers. Whilst on the seas his great protagonist Lavoisier met his death on the scaffold.

Such was the treatment bestowed upon the best of their citizens by two nations which considered themselves as without exception the most civilized and enlightened in the world!

It is quite natural to query how the grand old scientist busied himself on this voyage of eight weeks and a day. The answer is found in his own words:

I read the whole of the Greek Testament, and the Hebrew bible as far as the first Book of Samuel: also Ovid's Metamorphoses, Buchanan's poems, Erasmus' Dialogues, also Peter Pindar's poems, &c.... and [Pg 13] to amuse myself I tried the heat of the water at different depths, and made other observations, which suggest various experiments, which I shall prosecute whenever I get my apparatus at liberty.

The Doctor was quite sea-sick, and at times sad, but uplifted when his eyes beheld the proofs of friendship among those he was leaving behind. Thus he must have smiled benignantly on beholding the

elegant Silver Inkstand, with the following inscription, presented ... by three young Gentlemen of the University of Cambridge:

"To Joseph Priestley, LL.D. &c. on his departure into Exile, from a few members of the University of Cambridge, who regret that expression of their Esteem should be occasioned by the ingratitude of their Country."

And, surely, he must have taken renewed courage on perusing the valedictory message received from the Society of United Irishmen of Dublin: [Pg 14]

Sir,

SUFFER a Society which has been caluminated as devoid of all sense of religion, law or morality, to sympathize with one whom calumny of a similar kind is about to drive from his native land, a land which he has adorned and enlightened in almost every branch of liberal literature, and of useful philosophy. The emigration of Dr. Priestley will form a striking historical fact, by which alone, future ages will learn to estimate truly the temper of the present time. Your departure will not only give evidence of the injury which philosophy and literature have received in your person, but will prove the accumulation of petty disquietudes, which has robbed your life of its zest and enjoyment, for, at your age no one would willingly embark on such a voyage, and sure we are, it was your wish and prayer to be buried in your native country, which contains the dust of your old friends Saville, Price, Jebb, and Fothergill. But be cheerful, dear Sir, you are going to a happier world—the world of Washington and Franklin.

In idea, we accompany you. We stand near you while you are setting sail. We watch [Pg 15] your eyes that linger on the white cliffs and we hear the patriarchal blessing which your soul pours out on the land of your nativity, the aspiration that ascends to God for its peace, its freedom and its prosperity. Again, do we participate in your feelings on first beholding Nature in her noblest scenes and grandest features, on finding man busied in rendering himself worthy of Nature, but more than all, on contemplating with philosophic prescience the coming period when those vast inland seas shall be shadowed with sails, when the St. Lawrence and Mississippi, shall stretch forth their arms to embrace the continent in a great circle of interior navigation: when the Pacific Ocean shall pour into the Atlantic; when man will become more precious than fine gold, and when his ambition will be to subdue the elements, not to subjugate his fellow-creatures, to make fire, water, earth and air obey his bidding, but to leave the poor ethereal mind as the sole thing in Nature free and incoercible.

Happy indeed would it be were men in power to recollect this quality of the human mind. Suffer us to give them an example from a science of which you are a mighty [Pg 16] master, that attempts to fix the element of mind only increase its activity, and that to calculate what may be from what has been is a very dangerous deceit.— Were all the saltpetre in India monopolized, this would only make chemical researches more ardent and successful. The chalky earths would be searched for it, and nitre beds would be made in every cellar and every stable. Did not that prove sufficient the genius of chemistry would find in a new salt a substitute for nitre or a power superior to it. [3] It requires greater genius than Mr. Pitt seems to possess, to know the wonderful resources of the mind, when patriotism animates philosophy, and all the arts and sciences are put under a state of requisition, when the attention of a whole scientific people is bent to multiplying the means and instruments of destruction and when philosophy rises in a mass to drive on the wedge of war. A black powder has changed the military art, and in a great degree the manners of mankind. Why may not the same science which produced it, produce another powder which, inflamed under a certain compression, might impell the air, so as to shake down the strongest towers and scatter destruction. [Pg 17]

But you are going to a country where science is turned to better uses. Your change of place will give room for the matchless activity of your genius; and you will take a sublime pleasure in bestowing on Britain the benefit of your future discoveries. As matter changes its form but not a particle is ever lost, so the principles of virtuous minds are equally imperishable; and your change of situation may even render truth more operative, knowledge more productive, and in the event, liberty itself more universal. Wafted by the winds or tossed by the waves, the seed that is here thrown out as dead, there shoots up and flourishes. It is probable that emigration to America from the first settlement downward, has not only served the cause of general liberty, but will eventually and circuitously serve it even in Britain. What mighty events have arisen from that germ which might once have been supposed to be lost forever in the woods of America, but thrown upon the bosom of Nature, the [Pg 18] breath of God revived it, and the world hath gathered its fruits. Even Ireland has contributed her share to the liberties of America; and while purblind statesmen were happy to get rid of the stubborn Presbyterians of the North, they little thought that they were serving a good cause in another quarter. — Yes! the Volunteers of Ireland still live — they live across the Atlantic. Let this idea animate us in our sufferings, and may the pure principles and genuine lustre of the British Constitution reflected from their Coast, penetrate into ourselves and our dungeons.

Farewell—great and good man! Great by your mental powers, by your multiplied literary labours, but still greater by those household virtues which form the only solid security for public conduct by those mild and gentle qualities, which far from being averse to, are most frequently attended with severe and inflexible patriotism, rising like an oak above a modest mansion.—Farewell—but before you go, we beseech a portion of your parting prayer to the author of Good for Archibald Hamilton Rowan, the pupil of Jebb, our Brother, now suffering imprisonment, and for all those [Pg 19] who have suffered, and are about to suffer in the same cause—the cause of impartial and adequate representation—the cause of the Constitution. Pray to the best of Beings for Muir, Palmer, Skirving, Margarott and Gerald, who are now, or will shortly be crossing, like you, the bleak Ocean, to a barbarous land! — Pray that they may be animated with the same spirit, which in the days of their fathers, triumphed at the stake, and shone in the midst of flames. Melancholy indeed, it is that the mildest and most humane of all Religions should have been so perverted as to hang or burn men in order to keep them of one faith.

It is equally melancholy, that the most deservedly extolled of Civil Constitutions, should recur to similar modes of coercion, and that hanging and burning are not now employed, principally, because measures apparently milder are considered as more effectual. Farewell! Soon may you embrace your sons on the American shore, and Washington take you by the hand, and the shade of Franklin look down with calm delight on the first statesman of the age extending his protection to its first philosopher.

[Pg 20]

And how interestedly did America anticipate the arrival of the world renowned philosopher is in a measure foreshadowed by the following excerpt from the *American Daily Advertiser* for Thursday, June 5, 1794:

Dr. Priestley, with about one hundred other passengers, are on board the Sansom, which may be hourly expected.

In an editorial of the same paper, printed about the same date, there appeared the following tribute:

It must afford the most sincere gratification to every well wisher to the rights of man, that the United States of America, the land of freedom and independence, has become the asylum of the greatest characters of the present age, who have been persecuted in Europe, merely because they have defended the rights of the enslaved nations.

The name of Joseph Priestley will be long remembered among all enlightened people; and there is no doubt that England will one day regret her ungrateful treatment to this venerable and illustrious man. His [Pg 21] persecutions in England have presented to him the American Republic as a safe and honourable retreat in his declining years; and his arrival in this City calls upon us to testify our respect and esteem for a man whose whole life has been devoted to the sacred duty of diffusing knowledge and happiness among nations.

The citizens of united America know well the honourable distinction that is due to virtue and talents; and while they cherish in their hearts the memory of Dr. Franklin, as a philosopher, they will be proud to rank among the list of their illustrious fellow citizens, the name of Dr. Priestley.

Quietly but with great inward rejoicing were the travel-worn voyagers—the Doctor and his wife—received on the evening of June 4, 1794, at the old Battery in New York, by their son Joseph and his wife, who had long awaited them, and now conducted them to a nearby lodging house, which had been the head-quarters of Generals Howe and Clinton.

On the following morning the Priestleys were visited by Governor Clinton, Dr. Prevost, Bishop of New York and most of the principal merchants, and deputations of corporate bodies and Societies, [Pg 22] bringing addresses of welcome. Thus, among the very first to present their sympathetic welcome was the Democratic Society of the City of New York, which in the address of its President, Mr. James Nicholson, made June 7, 1794, said:

Sir,

WE are appointed by the Democratic Society of the City of New York, a Committee to congratulate you on your arrival in this country: And we feel the most lively pleasure in bidding you a hearty welcome to these shores of Liberty and Equality.

While the arm of Tyranny is extended in most of the nations of the world, to crush the spirit of liberty, and bind in chains the bodies and minds of men, we acknowledge, with ardent gratitude to the Great Parent of the Universe, our singular felicity in living in a land, where Reason has successfully triumphed over the artificial distinctions of European policy and bigotry, and where the law equally protects the virtuous citizen of every description and persuasion.

On this occasion we cannot but observe, that we once esteemed ourselves happy in the [Pg 23] relation that subsisted between us and the Government of Great Britain—But the multiplied oppressions which characterized that Government, excite in us the most painful sensations, and exhibit a spectacle as disgusting in itself, as dishonourable to the British name.

The governments of the old world present to us one huge mass of intrigue, corruption and despotism — most of them are now basely combined, to prevent the establishment of liberty in France, and to affect the total destruction of the rights of man. Under these afflicting circumstances we rejoice that America opens her arms to receive, with fraternal affection, the friend of liberty and human happiness, and that here he may enjoy the best blessings of civilized society.

We sincerely sympathize with you in all that you have suffered, and we consider the persecution with which you have been pursued by a venal Court and an imperious and uncharitable priesthood, as an illustrious proof of your personal merit, and a lasting reproach to that Government from the grasp of whose tyranny you are so happily removed. [Pg 24]

Accept, Sir, of the sincere and best wishes of the Society whom we represent, for the continuance of your health, and the increase of your individual and domestic happiness.

To which Priestley graciously replied:

Gentlemen,

VIEWING with the deepest concern, as you do, the prospect that is now exhibited in Europe, those troubles which are the natural offspring of their forms of government originating, indeed, in the spirit of liberty, but gradually degenerating in tyrannies, equally degrading to the rulers and the ruled, I rejoice in finding an asylum from persecution in a country in which these abuses have come to a natural termination, and have produced another system of liberty founded on such wise principles, as, I trust, will guard it against all future abuses; those artificial distinctions in society, from which they sprung, being completely eradicated, that protection from violence which laws and government promise in all countries, but which I have not found in my own, I doubt not I shall find with you, though, I cannot promise to be a better subject of this government, than [Pg 25] my whole conduct will evince that I have been to that of great Britain.

Justly, however, as I think I may complain of the treatment I have met with in England I sincerely wish her prosperity, and, from the good will I bear both that country and this I ardently wish that all former animosities may be forgotten and that a perpetual friendship may subsist between them.

And on Monday, June, 11, 1794, having taken the first opportunity to visit Priestley, the Tammany Society presented this address:

Sir,

A numerous body of freemen who associate to cultivate among them the love of liberty and the enjoyment of the happy Republican government under which they live and who for several years have been known in this city, by the name of the Tammany Society have deputed us a Committee to express to you their pleasure and congratulations on your safe arrival in this country.

Their venerable ancestors escaped, as you have done, from persecutions of intolerance, [Pg 26] bigotry and despotism, and they would deem themselves, an unworthy progeny were they not highly interested in your safety and happiness.

It is not alone because your various useful publications evince a life devoted to literature and the industrious pursuit of knowledge; not only because your numerous discoveries in Nature are so efficient to the progression of human happiness: but they have long known you to be the friend of mankind and in defiance of calumny and malice, an asserter of the rights of conscience and the champion of civil and religious liberty.

They have learned with regret and indignation the abandoned proceedings of those spoilers who destroyed your house and goods, ruined your philosophical apparatus and library, committed to the flames your manuscripts, pryed into the secrets of your private papers, and in their barbarian fury put your life itself in danger. They heard you also with exalted benevolence return unto them "blessings for curses:" and while you thus exemplified the undaunted integrity of the patriot, the mild and forbearing virtues of the Christian, [Pg 27] they hailed you victor in this magnanimous triumph over your enemies.

You have fled from the rude arm of violence, from the flames of bigotry, from the rod of lawless power: and you shall find refuge in the bosom of freedom, of peace, and of Americans.

You have left your native land, a country doubtless ever dear to you—a country for whose improvement in virtue and knowledge you have long disinterestedly laboured, for which its rewards are ingratitude, injustice and banishment. A country although now presenting a prospect frightful to the eyes of humanity, yet once the nurse of science, of arts, of heroes, and of freeman—a country which although at present apparently self devoted to destruction, we fondly hope may yet tread back the steps of infamy and ruin, and once more rise conspicuous among the free nations of the earth. In this advanced period of your life, when nature demands the sweets of tranquility, you have been constrained to encounter the tempestous deep, to risk disappointed prospects in a foreign land, to give up the satisfaction of domestic quiet, to tear yourself from the [Pg 28] friends of your youth, from a numerous acquaintance who revere and love you, and will long deplore your loss.

We enter, Sir, with emotion and sympathy into the numerous sacrifices you must have made, to an undertaking which so eminently exhibits our country as an asylum for the persecuted and oppressed, and into those regretful sensibilities your heart experienced when the shores of your native land were lessening to your view.

Alive to the impressions of this occasion we give you a warm and hearty welcome into these United States. We trust a country worthy of you; where Providence has unfolded a scene as new as it is august, as felicitating as it is unexampled. The enjoyment of liberty with but one disgraceful exception, pervades every class of citizens. A catholic and sincere spirit of toleration regulates society which rises into zeal when the sacred rights of humanity are invaded. And there exists a sentiment of free and candid inquiry which disdains shackles of tradition, promising a rich harvest of improvement and the glorious triumphs of truth. We hope, Sir, that the Great Being whose laws and works you [Pg 29] have made the study of your life, will smile upon and bless you—restore you to every domestic and philosophical enjoyment, prosper you in every undertaking, beneficial to mankind, render you, as you have been to your own, the ornament of this country, and crown you at last with immortal felicity and honour.

And to this the venerable scientist was pleased to say:

Gentlemen,

I think myself greatly honoured, flying as I do, from ill treatment in my native country, on account of my attachment to the cause of civil and religious liberty, to be received with the congratulations of "a Society of Freemen associated to cultivate the love of liberty, and the enjoyment of a happy Republican government." Happy would our venerable ancestors, as you justly call them, have been, to have found America such a retreat for them as it is to me, when they were driven hither; but happy has it proved to me, and happy will it be for the world, that in the wise and benevolent order of Providence, abuses of [Pg 30] power are ever destructive of itself, and favourable to liberty. Their strenuous exertions and yours now give me that asylum which at my time of life is peculiarly grateful to me, who only wish to continue unmolested those pursuits of various literature to which, without having ever entered into any political connexions my life has been devoted.

I join you in viewing with regret the unfavourable prospect of Great Britain formerly, as you say, the nurse of science, and of freemen, and wish with you, that the unhappy delusion that country is now under may soon vanish, and that whatever be the form of its government it may vie with this country in everything that is favourable to the best interests of mankind, and join with you in removing that only disgraceful circumstance, which you justly acknowledge to be an exception to the enjoyment of equal liberty, among yourselves. That the Great Being whose providence extends alike to all the human race, and to whose disposal I cheerfully commit myself, may establish whatever is good, and remove whatever is imperfect from your government and from every govern [Pg 31] ment in the known world, is the earnest prayer of,

Gentlemen,

Your respectful humble servant.

As Priestley had ever gloried in the fact that he was a teacher, what more appropriate in this period of congratulatory welcome, could have come to him than the following message of New York's teaching body:

The associated Teachers in the city of New York beg leave to offer you a sincere and hearty welcome to this land of tranquility and freedom.

Impressed with the idea of the real importance of so valuable an acquisition to the growing interests of science and literature, in this country, we are particularly happy that the honour of your first reception, has fallen to this state, and to the city of New York.

As labourers in those fields which you have occupied with the most distinguished eminence, at the arduous and important task of cultivating the human mind, we contemplate with peculiar satisfaction the auspicious influence which your personal [Pg 32] residence in this country, will add to that of your highly valuable scientific and literary productions, by which we have already been materially benefited.

We beg leave to anticipate the happiness of sharing in some degree, that patronage of science and literature, which it has ever been your delight to afford. This will give facility to our expressions; direct and encourage us in our arduous employments; assist us to form the man, and thereby give efficacy to the diffusion of useful knowledge.

Our most ardent wishes attend you, good Sir, that you may find in this land a virtuous simplicity, a happy recess from the intriguing politics and vitiating refinements of the European world. That your patriotic virtues may add to the vigour of our happy Constitution and that the blessings of this country may be abundantly remunerated into your person and your family.

And we rejoice in believing, that the Parent of Nature, by those secret communications of happiness with which he never fails to reward the virtuous mind, will here convey to you that consolation, support, and joy, which are independent of local circum [Pg 33] stances, and "Which the world can neither give nor take away."

Touched, indeed was Priestley by this simple, outspoken greeting from those who appreciated his genuine interest in the cause of education. Hence his reply was in a kindred spirit:

A welcome to this country from my fellow labourers in the instruction of youth, is, I assure you, peculiarly grateful to me. Classes of men, as well as individuals, are apt to form too high ideas of their own importance; but certainly one of the most important is, that which contributes so much as ours do to the cummunication of useful knowledge, as forming the characters of men, thereby fitting them for their several stations in society. In some form or other this has been my employment and delight; and my principal object in flying for an asylum to this country, "a land," as I hope you justly term it, "of virtuous simplicity, and a recess from the intriguing politics, and vicious refinements of the European world," is that I may, without molestation, pursue my favourite studies. And if I had an opportunity of making choice of an [Pg 34] employment for what remains of active exertion in life, it would be one in which I should as I hope I have hitherto done, contribute with you, to advance the cause of science, of virtue, and of religion.

Further, The Medical Society of the State of New York through Dr. John Charlton, its President, said:

PERMIT us, Sir, to wait upon you with an offering of our sincere congratulations, on your safe arrival, with your lady and family in this happy country, and to express our real joy, in receiving among us, a gentleman, whose labours have contributed so much to the diffusion and establishment of civil and religious liberty, and whose deep researches into the true principles of natural philosophy, have derived so much improvement and real benefit, not only to the sciences of chemistry and medicine, but to various other arts, all of which are necessary to the ornament and utility of human life.

May you, Sir, possess and enjoy, here, uninterrupted contentment and happiness, and may your valuable life be continued a farther blessing to mankind.

[Pg 35]

And in his answer Dr. Priestley remarked:

I THINK myself greatly honoured in being congratulated on my arrival in this country by a Society of persons whose studies bear some relation to my own. To continue, without fear of molestation, on account of the most open profession of any sentiments, civil or religious, those pursuits which you are sensible have for their object the advantage of all mankind, (being, as you justly observe, "necessary to the ornament and utility of human life") is my principal motive for leaving a country in which that tranquility and sense of security which scientificial pursuits require, cannot be had; and I am happy to find here, persons who are engaged in the same pursuits, and who have the just sense that you discover of their truly enviable situation.

As a climax to greetings extended in the City of New York, The Republican Natives of Great Britain and Ireland resident in that city said,

WE, the Republican natives of Great Britain and Ireland, resident in the city of New York, embrace, with the highest satisfac [Pg 36] tion, the opportunity which your arrival in this city presents, of bearing our testimony to your character and virtue and of expressing our joy that you come among us in circumstances of such good health and spirits.

We have beheld with the keenest sensibility, the unparallelled persecutions which attended you in your native country, and have sympathized with you under all their variety and extent. In the firm hope, that you are now completely removed from the effects of every species of intolerance, we most sincerely congratulate you.

After a fruitless opposition to a corrupt and tyrannical government, many of us have, like you, sought freedom and protection in the United States of America; but to this we have all been principally induced, from the full persuasion, that a republican representative government, was not merely best adapted to promote human happiness, but that it is the only rational system worthy the wisdom of man to project, or to which his reason should assent.

Participating in the many blessings which the government of this country is calculated to [Pg 37] insure, we are happy in giving it this proof of our respectful attachment: — We are only grieved, that a system of such beauty and excellence, should be at all tarnished by the existence of slavery in any form; but as friends to the Equal Rights of Man, we must be permitted to say, that we wish these Rights extended to every human being, be his complexion what it may. We, however, look forward with pleasing anticipation to a yet more perfect state of society; and, from that love of liberty which forms so distinguishing a trait in American character, are taught to hope that this last—this worse disgrace to a free government, will finally and forever be done away.

While we look back on our native country with emotions of pity and indignation at the outrages which humanity has sustained in the persons of the virtuous Muir, and his patriotic associates; and deeply lament the fatal apathy into which our countrymen have fallen; we desire to be thankful to the Great Author of our being that we are in America, and that it has pleased Him, in his Wise Providence, to make the United States an asylum not only from the immedi [Pg 38] ate tyranny of the British Government, but also from those impending calamities, which its increasing despotism and multiplied iniquities, must infallibly bring down on a deluded and oppressed people.

Accept, Sir, of our affectionate and best wishes for a long continuance of your health and happiness.

The answer of the aged philosopher to this address was:

I think myself peculiarly happy in finding in this country so many persons of sentiments similar to my own, some of whom have probably left Great Britain or Ireland on the same account, and to be so cheerfully welcomed by them on my arrival. You have already had experience of the difference between the governments of the two countries, and I doubt not, have seen sufficient reason to give the decided preference that you do to that of this. There all liberty of speech and of the press as far as politics are concerned, is at an end, and a spirit of intolerance in matters of religion is almost as high as in the time of the [Pg 39] Stuarts. Here, having no countenance from government, whatever may remain of this spirit, from the ignorance and consequent bigotry, of former times, it may be expected soon to die away; and on all subjects whatever, every man enjoys invaluable liberty of speaking and writing whatever he pleases.

The wisdom and happiness of Republican governments and the evils resulting from hereditary monarchical ones, cannot appear in a stronger light to you than they do to me. We need only look to the present state of Europe and of America, to be fully satisfied in this respect. The former will easily reform themselves, and among other improvements, I am persuaded, will be the removal of that vestige of servitude to which you allude, as it so ill accords with the spirit of equal liberty, from which the rest of the system has flowed; whereas no material reformation of the many abuses to which the latter are subject, it is to be feared, can be made without violence and confusion.

I congratulate you, gentlemen, as you do me, on our arrival in a country in which men who wish well to their fellow citizens, and [Pg 40] use their best endeavours to render them the most important services, men who are an honour to human nature and to any country, are in no danger of being treated like the worst felons, as is now the case in Great Britain.

Happy should I think myself in joining with you in welcoming to this country every friend of liberty, who is exposed to danger from the tyranny of the British Government, and who, while they continue under it, must expect to share in those calamities, which its present infatuation must, sooner or later, bring upon it. But let us all join in supplications to the Great Parent of the Universe, that for the sake of the many excellent characters in our native country its government may be reformed, and the judgments impending over it prevented.

The hearty reception accorded Dr. Priestley met in due course with a cruel attack upon him by William Cobbett, known under the pen-name of Peter Porcupine, an Englishman, who after arrival in this country enjoyed a rather prosperous life by formulating scurrilous literature—attacks upon men of prominence, stars shining brightly in the human firmament. [Pg 41]

An old paper, the *Argus*, for the year 1796, said of this Peter Porcupine:

When this political caterpillar was crawling about at St. John's, Nova Scotia, in support of his Britannic Majesty's glorious cause, against the United States, and holding the rank of serjeant major in the 54th regiment, then quartered in that land, "flowing with milk and honey," and GRINDSTONES, and commanded by Colonel Bruce; it was customary for some of the officers to hire out the soldiers to the country people, instead of keeping them to military duty, and to pocket the money themselves. Peter found he could make a *speck* out of this, and therefore kept a watchful eye over the sins of his superiors. When the regiment was recalled and had returned to England—Peter, brimful of amor patriæ, was about to prefer a complaint against the officers, when they came down with a round sum of the ready rino, and a promise of his discharge, in case of secrecy.—This so staggered our incorruptible and independent hero and quill driver, that he agreed to the terms, received that very honorable discharge, mentioned with so [Pg 42] much emphasis, in the history of his important life—got cash enough to come to America, by circuitous route and to set himself up with the necessary implements of scandal and abuse.

This flea, this spider, this corporal, has dared to point his impotent spleen at the memory of that illustrious patriot, statesman and philosopher, Benjamin Franklin.

Let the buzzing insect reflect on this truth—that

"Succeeding times great Franklin's works shall quote,
When 'tis forgot—this Peter ever wrote."

And the *Advertiser* declared:

Peter Porcupine is one of those writers who attempt to deal in wit—and to bear down every Republican principle by satire—but he miserably fails in both, for his wit is as stale as his satire, and his satire as insipid as his wit. He attempts to ridicule Dr. Franklin, but can any man of sense conceive any poignancy in styling this great philosopher, "poor Richard," or "the old lightning rod." Franklin, whose researches in philosophy have placed him preeminent [Pg 43] among the first characters in this country, or in Europe: is it possible then that such a contemptible wretch as Peter Porcupine, (who never gave any specimen of his philosophy, but in bearing with Christian patience a severe whipping at the public post) can injure the exalted reputation of this great philosopher? The folly of the Editor of the Centinal, is the more conspicuous, in inserting his billingsgate abuse in a Boston paper, when this town, particularly the TRADESMAN of it are reaping such advantages from Franklin's liberality. The Editor of the Centinal ought to blush for his arrogance in vilifying this TRADESMEN'S FRIEND, by retailing the scurrility of so wretched a puppy as Peter Porcupine.

As to Dr. Priestley, the Editor was obliged to apologise in this particular—but colours it over as the effusions of genius—poor apology, indeed to stain his columns with scurrility and abuse, and after finding the impression too notoriously infamous, attempts to qualify it, sycophantic parenthesis.

The names of Franklin and Priestley will be enrolled in the catalogue of worthies, while [Pg 44] the wretched Peter Porcupine, and his more wretched supporters, will sink into oblivion, unless the register of Newgate should be published, and their memories be raked from the loathsome rubbish as spectres of universal destestation.

And the London Monthly Review (August 10, 1796) commented as follows on Porcupine's animadversions upon Priestley:

Frequently as we have differed in opinion from Dr. Priestley, we should think it an act of injustice to his merit, not to say that the numerous and important services which he has rendered to science, and the unequivocal proofs which he has given of at least honest intention towards religion and Christianity ought to have protected him from such gross insults as are poured upon him in this pamphlet. Of the author's literary talent, we shall say but little: the phrases, "setting down to count the cost"—"the rights of the man the greatest bore in nature"—the appellation of rigmarole ramble, given to a correct sentence of Dr. Priestley—which the author attempts [Pg 45] to criticise—may serve as specimens of his language.

The pitiful attempt at wit, in his vulgar fable of the pitcher haranguing the pans and jordans, will give him little credit as a writer, with readers of an elegant taste.—No censure, however, can be too severe for a writer who suffers the rancour of party spirit to carry him so far beyond the bounds of justice, truth and decency, as to speak of Dr. Priestley as an admirer of the massacres of France, and who would have wished to have seen the town of Birmingham like that of Lyons, razed, and all its industrious and loyal inhabitants butchered as a man whose conduct proves that he has either an understanding little superior to that of an idiot, or the heart of Marat: in short, as a man who fled into banishment covered with the universal destestation of his countrymen. The spirit, which could dictate such outrageous abuse, must disgrace any individual and any party.

Even before Porcupine began his abuse of Priestley, there appeared efforts intended no doubt to arouse opposition to him and dislike for him. One such, apparently very innocent in its purpose, [Pg 46] appeared shortly after Priestley's settlement in Northumberland. It may be seen in *the Advertiser*, and reads thus:

The divinity of Jesus Christ proved in a publication to be sold by Francis Bayley in Market Street, between 3rd and 4th Streets, at the sign of the *Yorick's Head*—being a reply to Dr. Joseph Priestley's appeal to the serious and candid professors of Christianity.

The New York addresses clearly indicated the generous sympathy of hosts of Americans for Priestley. They were not perfunctory, but genuinely genuine. This brought joy to the distinguished emigrant, and a sense of fellowship, accompanied by a feeling of security.

More than a century has passed since these occurrences, and the reader of today is scarcely stirred by their declarations and appeals. Changes have come, in the past century, on both sides of the great ocean. Almost everywhere reigns the freedom so devoutly desired by the fathers of the long ago. It is so universal that it does not come as a first thought. Other changes, once constantly on men's minds have gradually been made. [Pg 47]

How wonderful has been the development of New York since Priestley's brief sojourn in it. How marvelously science has grown in the great interim. What would Priestley say could he now pass up and down the famous avenues of our greatest City?

His decision to live in America, his labors for science in this land, have had a share in the astounding unfolding of the dynamical possibilities of America's greatest municipality.

The Priestleys were delighted with New York. They were frequent dinner guests of Governor Clinton, whom they liked very much and saw often, and they met with pleasure Dr. Samuel L. Mitchill, the Professor of Chemistry in Columbia.

Amidst the endless fetes, attendant upon their arrival, there existed a desire to go forward. The entire family were eager to arrive at their real resting place—the home prepared by the sons who had preceded them to this Western world. Accordingly, on June 18, 1794, they left New York, after a fortnight's visit, and the *Advertiser* of Philadelphia, June 21, 1794, contained these lines:

Last Thursday evening arrived in town from New York the justly celebrated philosopher Dr. Joseph Priestley.

[Pg 48]

Thus was heralded his presence in the City of his esteemed, honored friend, Franklin, who, alas! was then in the spirit land, and not able to greet him as he would have done had he still been a living force in the City of Brotherly Love. However, a very prompt welcome came from the American Philosophical Society, founded (1727) by the immortal savant, Franklin.

The President of this venerable Society, the oldest scientific Society in the Western hemisphere, was the renowned astronomer, David Rittenhouse, who said for himself and his associates:

THE American Philosophical Society, held at Philadelphia for promoting useful knowledge, offer you their sincere congratulations on your safe arrival in this country. Associated for the purposes of extending and disseminating those improvements in the sciences and the arts, which most conduce to substantial happiness of Man, the Society felicitate themselves and their country, that your talents and virtues, have been transferred to this Republic. Considering you as an illustrious member of this institution: Your colleagues anticipate your aid, in zealously promoting the [Pg 49] objects which unite them; as a virtuous man, possessing eminent and useful acquirements, they contemplate with pleasure the accession of such worth to the American Commonwealth, and looking forward to your future character of a citizen of this, your adopted country, they rejoice in greeting, as such, an enlightened Republican.

In this free and happy country, those unalienable rights, which the Author of Nature committed to man as a sacred deposit, have been secured: Here, we have been enabled, under the favour of

Divine Providence, to establish a government of Laws, and not of Men; a government, which secures to its citizens equal Rights, and equal Liberty, and which offers an asylum to the good, to the persecuted, and to the oppressed of other climes.

May you long enjoy every blessing which an elevated and highly cultivated mind, a pure conscience, and a free country are capable of bestowing.

And, in return, Priestley remarked.

IT is with peculiar satisfaction that I receive the congratulations of my brethren of [Pg 50] the Philosophical Society in this City, on my arrival in this country. It is, in great part, for the sake of pursuing our common studies without molestation, though for the present you will allow, with far less advantage, that I left my native country, and have come to America; and a Society of Philosophers, who will have no objection to a person on account of his political or religious sentiments, will be as grateful, as it will be new to me. My past conduct, I hope, will show, that you may depend upon my zeal in promoting the valuable objects of your institution; but you must not flatter yourself, or me, with supposing, that, at my time of life, and with the inconvenience attending a new and uncertain settlement, I can be of much service to it.

I am confident, however, from what I have already seen of the spirit of the people of this country, that it will soon appear that Republican governments, in which every obstruction is removed to the exertion of all kinds of talent, will be far more favourable to science, and the arts, than any monarchical government has ever been. The patronage to be met with there is ever capricious, and as often employed to bear down [Pg 51] merit as to promote it, having for its real object, not science or anything useful to mankind, but the mere reputation of the patron, who is seldom any judge of science. Whereas a Public which neither flatters nor is to be flattered will not fail in due time to distinguish true merit and to give every encouragement that it is proper to be given in the case. Besides by opening as you generously do an asylum to the persecuted and "oppressed of all climes," you will in addition to your own native stock, soon receive a large accession of every kind of merit, philosophical not excepted, whereby you will do yourselves great honour and secure the most permanent advantage to the community.

Doubtless in the society of so many worthy Philadelphians, the Priestleys were happy, for they had corresponded with not a few of them.

The longing for Northumberland became very great and one smiles on reading that the good Doctor thought "Philadelphia by no means so agreeable as New York ... Philadelphia would be very irksome to me.... It is only a place for business and to get money in." But in this City he later spent much of his time. [Pg 52]

It was about the middle of July, 1794, that the journey to Northumberland began, and on September 14, 1794, Priestley wrote of Northumberland "nothing can be more delightful, or more healthy than this place."

Safely lodged among those dear to him one finds much pleasure in observing the great philosopher's activities. The preparation of a home for himself and his wife and the unmarried members of the family was uppermost in his mind. But much time was given to correspondence with loyal friends in England. Chief among these were the Reverends Lindsey and Belsham. The letters to these gentlemen disclose the plans and musings of the exile. For instance, in a communication to the former, dated September 14, 1794, he wrote:

The professor of chemistry in the College of Philadelphia is supposed to be on his death-bed ... in the case of a vacancy, Dr. Rush thinks I shall be invited to succeed him. In this case I must reside four months in one year in Philadelphia, and one principal inducement with me to accept of it will be the opportunity I shall have of forming an Unitarian Congregation....

[Pg 53]

And a month later he observed to the same friend:

Philadelphia is unpleasant, unhealthy, and intolerably expensive.... Every day I do something towards the continuation of my Church History.... I have never read so much Hebrew as I have since I left England....

He visited freely in the vicinity of Northumberland, spending much time in the open. Davy, a traveler, made this note:

Dr. Priestley visited us at Sunbury, looks well and cheerful, has left off his perriwig, and combs his short grey locks, in the true style of the simplicity of the country.... Dined very pleasantly with him. He has bought a lot of eleven acres (exclusively of that which he is building on), which commands a delightful view of all the rivers, and both towns, i.e. Sunbury and Northumberland and the country. It cost him 100£ currency.

It was also to Mr. Lindsey that he communicated, on November 12, 1794, a fact of no little [Pg 54] interest, even today, to teachers of Chemistry in America. It was:

I have just received an invitation to the professorship of chemistry at Philadelphia ... when I considered that I must pass four months of every year from home, my heart failed me; and I declined it. If my books and apparatus had been in Philadelphia, I might have acted differently, but part of them are now arrived here, and the remainder I expect in a few days, and the expense and risk of conveyance of such things from Philadelphia hither is so great, that I cannot think of taking them back ... and in a year or two, I doubt not, we shall have a college established here.

It was about this time that his youngest son, Harry, in whom he particularly delighted, began clearing 300 acres of cheap land, and in this work the philosopher was greatly interested; indeed, on occasions he actually participated in the labor of removing the timber. Despite this manual labor there were still hours of every day given to the Church History, and to his correspondence which grew in volume, as he was advising inquiring English friends, who [Pg 55] thought of emigrating, and very generally to them he recommended the perusal of Dr. Thomas Cooper's

"Advice to those who would remove to America—"

Through this correspondence, now and then, there appeared little animadversions on the quaint old town on the Delaware, such as

I never saw a town I liked less than Philadelphia.

Could this dislike have been due to the fact that—

33

Probably in no other place on the Continent was the love of bright colours and extravagance in dress carried to such an extreme. Large numbers of the Quakers yielded to it, and even the very strict ones carried gold-headed canes, gold snuff-boxes, and wore great silver buttons on their drab coats and handsome buckles on their shoes.

And

Nowhere were the women so resplendant in silks, satins, velvets, and brocades, and they piled up their hair mountains high.

[Pg 56]

Furthermore—

The descriptions of the banquets and feasts ... are appalling.

John Adams, when he first came down to Philadelphia, fresh from Boston, stood aghast at this life into which he was suddenly thrown and thought it must be sin. But he rose to the occasion, and, after describing in his diary some of the "mighty feasts" and "sinful feasts" ... says he drank Madeira "at a great rate and found no inconvenience."

It would only be surmise to state what were the Doctor's reasons for his frequent declaration of dislike for Philadelphia.

The winter of 1794-1795 proved much colder "than ever I knew it in England," but he cheerfully requested Samuel Parker to send him a hygrometer, shades or bell-glasses, jars for electrical batteries, and

a set of glass tubes with large bulbs at the end, such as I used in the experiments I last published on the generation of *Air* from water.

[Pg 57]

Most refreshing is this demand upon a friend. It indicates the keen desire in Priestley to proceed with experimental studies, though surroundings and provisions for such undertakings were quite unsatisfactory. The spirit was there and very determined was its possessor that his science pursuits should not be laid totally aside. His attitude and course in this particular were admirable and exemplary. Too often the lack of an abundance of equipment and the absence of many of the supposed essentials, have been deterrents which have caused men to abandon completely their scientific investigations. However, such was not the case with the distinguished exile, and for this he deserved all praise.

From time to time, in old papers and books of travel, brief notes concerning Priestley appear. These exhibit in a beautiful manner the human side of the man. They cause one to wish that the privilege of knowing this worthy student of chemical science might have been enjoyed by him. For example, a Mr. Bakewell chanced upon him in the spring of 1795 and recorded:

I found him (Priestley) a man rather below the middle size, straight and plain, wearing his own hair; and in his countenance, though you might discern the philosopher, [Pg 58] yet it beamed with so much simplicity and freedom as made him very easy of access.

It is also stated in Davy's "Journal of Voyage, etc."—

The doctor enjoys a game at whist; and although he never hazards a farthing, is highly diverted with playing good cards, but never ruffled by bad ones.

In May, 1795, Priestley expressed himself as follows:

As to the experiments, I find I cannot do much till I get my own house built. At present I have all my books and instruments in one room, in the house of my son.

This is the first time in all his correspondence that reference is made to experimental work. It was in 1795. As a matter of course every American chemist is interested to know when he began experimentation in this country.

In the absence of proper laboratory space and the requisite apparatus, it is not surprising that he thought much and wrote extensively on religious topics, and further he would throw himself into [Pg 59] political problems, for he addressed Mr. Adams on restriction "in the naturalization of foreigners." He remarked that—

Party strife is pretty high in this country, but the Constitution is such that it cannot do any harm.

To friends, probably reminding him of being "unactive, which affects me much," he answered:

As to the chemical lectureship (in Philadelphia) I am convinced I could not have acquitted myself in it to proper advantage. I had no difficulty in giving a general course of chemistry at Hackney (England), lecturing only once a week; but to give a lecture every day for four months, and to enter so particularly into the subject as a course of lectures in a medical University (Pennsylvania) requires, I was not prepared for; and my engagements there would not, at my time of life, have permitted me to make the necessary preparations for it; if I could have done it at all. For, though I have made discoveries in some branches of chemistry, I never gave much attention to the common routine of [Pg 60] it, and know but little of the common processes.

Is not this a refreshing confession from the celebrated discoverer of oxygen? The casual reader would not credit such a statement from one who August 1, 1774, introduced to the civilized world so important an element as oxygen. Because he did not know the "common processes" of chemistry and had not concerned himself with the "common routine" of it, led to his blazing the way among chemical compounds in his own fashion. Many times since the days of Priestley real researchers after truth have proceeded without compass and uncovered most astonishing and remarkable results. They had the genuine research spirit and were driven forward by it. Priestley knew little of the labyrinth of analysis and cared less; indeed, he possessed little beyond an insatiable desire to unfold Nature's secrets.

Admiration for Priestley increases on hearing him descant on the people about him — on the natives —

Here every house-keeper has a garden, out of which he raises almost all he wants for his family. They all have cows, and many have horses, the keeping of which costs [Pg 61] them little or nothing in the summer, for they ramble with bells on their necks in the woods, and come home at night. Almost all the fresh meat they have is salted in the autumn, and a fish called *shads* in the spring. This salt shad they eat at breakfast, with their tea and coffee, and also at night. We, however, have not yet laid aside our English customs, and having made great exertion to get fresh meat, it will soon come into general use.

Proudly must he have said —

My youngest son, Harry, works as hard as any farmer in the country and is as attentive to his farm, though he is only eighteen.... Two or three hours I always work in the fields along with my son....

And, then as a supplement, for it was resting heavily on his mind, he added —

What I chiefly attend to now is my Church History ... but I make some experiments every day (July 12, 1795), and shall soon draw up a paper for the Philosophical Society at Philadelphia.

[Pg 62]

Early in December of 1795 he entrusted a paper, intended for the American Philosophical Society to the keeping of Dr. Young, a gentleman from Northumberland en route for Europe. Acquainting his friend Lindsey of this fact, he took occasion to add —

I have much more to do in my laboratory, but I am under the necessity of shutting up for the winter, as the frost will make it impossible to keep my water fit for use, without such provision as I cannot make, till I get my own laboratory prepared on purpose, when I hope to be able to work alike, winter and summer.

Dr. Young carried two papers to Philadelphia. The first article treated of "Experiments and Observations relating to the Analysis of Atmospherical Air," and the second "Further Experiments relating to the Generation of Air from Water." They filled 20 quarto pages of Volume 4 of the Transactions of the American Philosophical Society. On reading them the thought lingers that these are the first contributions of the eminent philosopher from his American home. Hence, without reference to their value, they are precious. [Pg 63] They represent the results of inquiries performed under unusual surroundings. It is very probable that Priestley's English correspondents desired him to concentrate his efforts upon experimental science. They were indeed pleased to be informed of his Church History, and his vital interest in religion, but they cherished the hope that science would in largest measure displace these literary endeavors. Priestley himself never admitted this, but must have penetrated their designs, and, recognizing the point of their urging, worked at much disadvantage to get the results presented in these two pioneer studies. Present day students would grow impatient in their perusal, because of the persistent emphasis placed on phlogiston, dephlogisticated air, phlogisticated air, and so forth. In the very first paper, the opening lines show this:

It is an essential part of the antiphlogistic theory, that in all the cases of what I have called *phlogistication* of *air*, there is simply an absorption of the dephlogisticated air, or, as the advocates of that theory term it, the oxygen contained in it, leaving the *phlogisticated* part, which they call *azote*, as it originally existed in the atmosphere. Also, according to this system, *azote* is a simple [Pg 64] substance, at least not hitherto analyzed into any other.

No matter how deeply one venerates Priestley, or how great honor is ascribed to him, the question continues why the simpler French view was not adopted by this honest student. Further, as an ardent admirer one asks why should Priestley pen the next sentence:

They, therefore, suppose that there is a determinate proportion between the quantities of oxygen, and azote in every portion of atmospherical air, and that all that has hitherto been done has been to separate them from one another. This proportion they state to be 27 parts of oxygen and 73 parts of azote, in 100 of atmospherical air.

Priestley knew that there was a "determinate proportion." He was not, however, influenced by quantitative data.

Sir Oliver Lodge said [4] —

Priestley's experiments were admirable, but his perception of their theoretical relations [Pg 65] was entirely inadequate and, as we now think, quite erroneous.... In theory he had no instinct for guessing right ... he may almost be said to have had a predilection for the wrong end.

At present the French thought is so evident that it seems incomprehensible that Priestley failed to grasp it, for he continues —

In every case of the diminution of atmospherical air in which this is the result, there appears to me to be something emitted from the substance, which the antiphlogistians suppose to act by simple absorption, and therefore that it is more probable that there is some substance, and the same that has been called *philogiston* , or the *prin-*

ciple of inflammability ... emitted, and that this phlogiston uniting with part of the dephlogisticated air forms with it part of the phlogisticated air, which is found after the process.

Subsequently (1798), he advised the Society that he had executed other experiments which corroborated those outlined in his first two papers, adding [Pg 66] —

Had the publication of your *Transactions* been more frequent, I should with much pleasure have submitted to the Society a full account of these and other experiments which appear to me to prove, that metals are compound substances, and that water has not yet been decomposed by any process that we are acquainted with. Still, however, I would not be very positive, as the contrary is maintained by almost all the chemists of the age....

And thus he proceeds, ever doing interesting things, but blind to the patent results because he had phlogiston constantly before him. He looked everywhere for it, followed it blindly, and consequently overlooked the facts regarded as most significant by his opponents, which in the end led them to correct conclusions.

The experimental results in the second paper also admit of an interpretation quite the opposite of that deduced by Priestley. He confidently maintained that air was invariably generated from water, because he discovered it and liberated it from water which he was certain did not contain it in solution. He was conscientious in his inferences. Deeply did his friends deplore his [Pg 67] inability to see more than a single interpretation of his results!

The papers were read before the American Philosophical Society on the 19th of February, 1796. Their author as they appear in print, is the Rev. Dr. J. Priestley. It is doubtful whether he affixed this signature. More probable is it that the Secretary of the Society was responsible, and, because he thought of Priestley in the rôle of a Reverend gentleman rather than as a scientific investigator.

Here, perhaps, it may be mentioned that the first, the very first communication from Priestley's pen to the venerable Philosophical Society, was read in 1784. It was presented by a friend—a Mr. W. Vaughan, whose family in England were always the staunchest of Priestley's supporters. And it is not too much to assume that it was the same influence which one year later (1785) brought about Priestley's election to membership in the Society, for he was one of "28 new members" chosen in January of that year.

There are evidences of marked friendliness to Priestley all about the Hall of the Society, for example his profile in Plaster of Paris, "particularly valuable for the resemblance" to the Doctor, which was presented in 1791; a second "profile in black leather" given by Robert Patterson, [Pg 68] a President of the Society, and an oil portrait of him from Mrs. Dr. Caspar Wistar.

His appearance in person, when for the first time he sat among his colleagues of the Society, was on the evening of February 19, 1796 — the night upon which the two papers, commented upon in the last few paragraphs were presented, although he probably did not read them himself, this being done by a friend or by the secretary. Sixteen members were present. Among these were some whose names have become familiar elsewhere, such as Barton, Woodhouse and others. Today, the presence in the same old Hall of a renowned scientist, from beyond the seas, would literally attract crowds. Then it was not the fashion. But probably he had come unannounced and unheralded. Further, he was speaking at other hours on other topics in the city.

It is not recorded that he spoke before the philosophers. Perhaps he quietly absorbed their remarks and studied them, although he no doubt was agreeably aroused when Mr. Peale presented

to the Society a young son of four months and four days old, being the first child born in the Philosophical Hall, and requested that the Society would give him a name. On which the Society unanimously [Pg 69] agreed that, after the name of the chief founder and late President of the Society, he should be called Franklin.

In anticipation of any later allusion to Priestley's sojourn in Philadelphia be it observed that he attended meetings of the American Philosophical Society three times in 1796, twice in 1797, three times in 1801 and once in 1803, and that on February 3rd, 1797, he was chosen to deliver the annual oration before the Society, but the Committee reported that they waited on Dr. Priestley last Monday afternoon, who received the information with great politeness, but declined accepting of the appointment.

This lengthy digression must now be interrupted. It has gone almost too far, yet it was necessary in order that an account of the early experimental contributions of the exile might be introduced chronologically. As already remarked, Americans are most deeply interested in everything Priestley did during his life in this country and particularly in his scientific activities.

On resuming the story of the routine at Northumberland in the closing months of the year [Pg 70] 1795, there comes the cry from an agonized heart, —

We have lost poor Harry!

This was the message to a Philadelphia resident—a friend from old England. The loss, for such it emphatically was, affected the Doctor and Mrs. Priestley very deeply. This particular son was a pride to them and though only eighteen years old had conducted his farm as if he had been bred a farmer.

He was uncommonly beloved by all that worked under him.

His home was just outside of the borough of Northumberland. It was the gift of his father. His interment in "a plot of ground" belonging to the Society of Friends is thus described by Mr Bakewell:

I attended the funeral to the lonely spot, and there I saw the good old father perform the service over the grave of his son. It was an affecting sight, but he went through it with fortitude, and after praying, addressed the attendants in a few words, [Pg 71] assuring them that though death had separated them here, they should meet again in another and a better life.

The correspondence to friends in England was replete with accounts of lectures which were in process of preparation. They were discourses on the Evidences of Revelation and their author was most desirous of getting to Philadelphia that he might there deliver them. At that time this City was full of atheism and agnosticism. Then, too, the hope of establishing a Unitarian Church was ever in

Priestley's thoughts. How delightful it is to read, February 12th, 1796—

I am now on my way to Philadelphia.

When he left it in 1794 he was rather critical of it, but now after three days he arrived there. It was

a very good journey, accompanied by my daughter-in-law, in my son's Yarmouth waggon, which by means of a seat constructed of straw, was very easy.

Yes, back again to the City which was the only city in this country ever visited by him. Although at times he considered going to New York, [Pg 72] and even to Boston, Philadelphia was to become his Mecca. In it he was to meet the most congenial scientific spirits, and to the younger of these he was destined to impart a new inspiration for science, and for chemical science in particular. At the close of the three days' journey he wrote—

I am a guest with Mr. Russell.... We found him engaged to drink tea with President Washington, where we accompanied him and spent two hours as in any private family. He (Washington) invited me to come at any time, without ceremony. Everything is the reverse of what it is with you.

This was his first meeting with Washington. The spirit of the occasion impressed him. The democratic behavior of the great Federalist must have astonished him, if he ever entertained, as Lord Brougham would have us believe, a hostile opinion and thought him ungrateful because he would not consent to make America dependent upon France.

Priestley's eagerness to preach was intense, and happy must he have been on the day following his arrival, when his heart's wish was gratified. He preached in the church of Mr. Winchester [Pg 73] —

to a very numerous, respectable, and very attentive audience.

Many were members of Congress, and according to one witness—

The Congregation that attended were so numerous that the house could not contain them, so that as many were obliged to stand as sit, and even the doorways were crowded with people. Mr. Vice-President Adams was among the regular attendants.

All this greatly encouraged the Doctor. His expectations for the establishment of a Unitarian congregation were most encouraging. He declared himself ready to officiate every winter without salary if he could lodge somewhere with a friend. The regular and punctual attendance of Mr. Adams pleased him so much that he resolved on printing his sermons, for they were in great demand, and to dedicate the same to the Vice-President. He was also gratified to note that the "violent prejudice" to him was gradually being overcome. Today we smile on recalling the reception accorded the good Doctor in his early days in Philadelphia. We smile and yet our hearts fail to understand just why he should [Pg 74] have been so ostracised. To confirm this it may be noted that on one occasion Priestley preached in a Presbyterian Chapel, very probably in Northumberland, when one of the ministers was so displeased—

that he declared if they permitted him any more, he would never enter the puplit again.

And in 1794 on coming the first time to Philadelphia he wrote

There is much jealousy and dread of me.

How shameful and yet it was most real. Bakewell narrates that

"I went several times to the Baptist meeting in Second Street, under the care of Dr. Rogers. This man burst out, and bade the people beware, for 'a Priestley had entered the land;' and then, crouching down in a worshiping attitude, exclaimed, 'Oh, Lamb of God! how would they pluck thee from thy throne!'"

The public prints flayed Rogers, and even the staid old Philosophical Society indicated to him [Pg 75] that such conduct ill became a member of that august body. Accordingly humiliated he repented his error and in time became strongly attached to Priestley, concerning whom he told this story to a Mr. Taylor whose language is here given:

The Doctor (Priestley) would occasionally call on Dr. Rogers, and without any formal invitation, pass an evening at his house. One afternoon he was there when Dr. Rogers was not at home, having been assured by Mrs. Rogers that her husband would soon be there. Meanwhile, Mr. — —, a Baptist minister, called on Dr. Rogers, and being a person of rough manners, Mrs. Rogers was a good deal concerned lest he should say something disrespectful to Dr. Priestley in case she introduced the Doctor to him. At last, however, she ventured to announce Dr. Priestley's name, who put out his hand; but instead of taking it the other immediately drew himself back, saying, as if astonished to meet with Dr. Priestley in the home of one of his brethren, and afraid of being contaminated by having any social intercourse with him, 'Dr. Priestley! I can't be cordial.' [Pg 76]

It is easy to imagine that by this speech Mrs. Rogers was greatly embarrassed. Dr. Priestley, observing this, instantly relieved her by saying, and with all that benevolent expression of countenance and pleasantness of manner for which he was remarkable, 'Well, well, Madam, you and I can be cordial; and Dr. Rogers will soon be with us, Mr. — — and he can converse together, so that we shall all be very comfortable.' Thus encouraged, Mrs. Rogers asked Dr. Priestley some questions relative to the Scripture prophecies, to which he made suitable replies; and before Dr. Rogers arrived, Mr. — — was listening with much attention, sometimes making a remark or putting in a question. The evening was passed in the greatest harmony, with no inclination on the part of Mr. — — to terminate the conversation. At last Dr. Priestley, pulling out his watch, informed Mr. — — that as it was *ten* o'clock it was time that two old men like them were at their quarters. The other at first was not willing to believe that Dr. Priestley's watch was accurate; but finding that it was correct, he took his leave with apparent regret, observing that he had never spent a shorter and more [Pg 77] pleasant evening. He then went away, Dr. Priestley accompanying him, until it became necesary to separate. Next morning he called on his friend, Dr. Rogers, when he made the following frank and manly declaration: 'You and I well know that Dr. Priestley is quite wrong in regard to his theology, but notwithstanding this, he is a great and good man, and I behaved to him at our first coming together like a fool and a brute.'

Many additional evidences might be introduced showing that the Doctor was slowly winning his way among the people. It must also be remembered that not all of his associates were of the clerical group but that he had hosts of scientists as sincere and warm supporters. In Woodhouse's laboratory he was ever welcome and there must have met many congenial spirits who never discussed politics or religion. This was after the manner of the Lunar Society in Birmingham in which representatives of almost every creed came together to think of scientific matters. Hence, it is quite probable that Priestley's visit to Philadelphia was on the whole full of pleasure.

He was also in habits of close intimacy with Dr. Ewing, Provost of the University of Pennsyl [Pg 78] vania, and with the Vice-Provost, Dr. John Andrews, as well as with Dr. Benjamin Rush who had long been his friend and with whom he corresponded at frequent intervals after his arrival in America. To him Priestley had confided his hope of getting a college in Northumberland and inquired,—

Would the State give any encouragement to it?

To Rush he also wrote excusing

my weakness (for such you will consider it) when, after giving you reason to expect that I would accept the professorship of Chemistry, if it was offered to me, I now inform you that I must decline it.

Now and then he also advised him of such experiments as he was able to do; for example—

I made trial of the air of Northumberland by the test of nitrous air, but found it not sensibly different from that of England.

In the leisure he enjoyed his figure was often seen in Congress. He relished the debates which at the time were on the Treaty with England. He [Pg 79] declared he heard as good speaking there as in the House of Commons. He observed—

A Mr. Amos speaks as well as Mr. Burke; but in general the speakers are more argumentative, and less rhetorical. And whereas there are with you not more than ten or a dozen tolerable speakers, here every member is capable of speaking.

While none of the letters to Priestley's friends mention a family event of some importance the *American Advertiser*, February 13, 1796, announced that

Mr. William Priestley, second son of the celebrated Dr. Joseph Priestley, was married to the agreeable Miss Peggy Foulke, a young lady possessed with every quality to render the marriage state happy.

This occurred very probably just before the Doctor set forth from Northumberland to make his first Philadelphia visit. It is singular that little is said of the son William by the Doctor. Could it be that, in some way, he may have offended his parent? In his *Memorial* Rush, writing in the month of March, 1796, noted: [Pg 80]

Saw Dr. Priestley often this month. Attended him in a severe pleurisy. He once in his sickness spoke of his second son, William, and wept very much.

Busy as he was in spreading his religious tenets, in fraternizing with congenial scientific friends, his thoughts would involuntarily turn back to England:

Here, though I am as happy as this country can make me ... I do not feel as I did in England.

By May, 1796, he had finished his discourses, although he proposed concluding with one emphatically Unitarian in character. This was expected by his audience, which had been quietly prepared for it and received it with open minds and much approval.

On his return to Northumberland he promptly resumed his work on the "Church History," but was much disturbed because of the failure of his correspondents in writing him regularly, so he became particularly active in addressing them. But better still he punctuated his composition of sermons, the gradual unfolding of his Church History, and religious and literary studies in gene [Pg 81] ral, with experimental diversions, beginning with the publication (1796) of an octavo brochure of 39 pages from the press of Dobson in Philadelphia, in which he addressed himself more especially to Berthollet, de la Place, Monge, Morveau, Fourcroy and others on "Considerations on the Doctrine of Phlogiston and the Decomposition of Water." It is the old story in a newer dress. Its purpose was to bring home to Americans afresh his particular ideas. The reviewer of the *Medical Repository* staff was evidently impressed by it, for he said:

It must give pleasure to every philosophical mind to find the United States becoming the theatre of such interesting discussion, and then adds that the evidence which was weighty enough to turn such men as Black and others from the phlogiston idea to that of Lavoisier—

has never yet appeared to Dr. Priestley considerable enough to influence his judgment, or gain his assent.

Priestley, as frequently observed, entertained grave doubts in regard to the constitution of metals. He thought they were "compounded" [Pg 82] of a certain earth, or calx, and phlogiston. Further he believed that when the phlogiston flew away, "the splendour, malleability, and ductility" of the metal disappeared with it, leaving behind a calx. Again, he contended that when metals dissolved in acids the liberated "inflammable air" (hydrogen) did not come from the 'decompounded water' but from the phlogiston emitted by the metal.

Also, on the matter of the composition and decomposition of water, he held very opposite ideas. The French School maintained "that hydrogenous and oxygenous airs, incorporated by drawing through them the electrical spark turn to *water*," but Priestley contended that "they combine into *smoking nitrous acid*." And thus the discussion proceeded, to be answered most intelligently, in 1797, by Adet, [5] whose arguments are familiar to all chemists and need not therefore be here repeated. Of more interest was the publication of two lectures on Combustion by Maclean of Princeton. They filled a pamphlet of 71 pages. It appeared in 1797, and was, in brief, a refutation of Priestley's presentations, and was heartily welcomed as evidence of the "growing taste in America for this kind of inquiry." Among other things [Pg 83] Maclean said of the various ideas regarding combustion—"Becker's is incomplete, Stahl's though ingenious, is defective; the antiphlogistic is simple, consistent and sufficient, while Priestley's resembling Stahl's but in name, is complicated, contradictory and inadequate."

Not all American chemists were ready to side track the explanations of Priestley. The distinguished Dr. Mitchill wrote Priestley on what he designated "an attempt to accommodate the Disputes among Chemists concerning Phlogiston." This was in November, 1797. It is an ingenious effort which elicited from Priestley (1798) his sincere thanks, and the expressed fear that his labours "will be in vain." And so it proved. Present day chemists would acquiesce in this statement after reading Mitchill's "middle-of-the-road" arguments. They were not satisfactory to Maclean and irritated Priestley.

In June 1798 a second letter was written by Priestley to Mitchill. In it he emphasized the substitution of zinc for "finery cinder." From it he contended inflammable air could be easily procured, and laid great stress on the fact that the "inflammable air" came from the metal and not from the water. He wondered why Berthollet and Maclean had not answered his first article. To this, a few days later, Mitchill replied that he [Pg 84] felt there was confusion in terms and that the language employed by the various writers had introduced that confusion; then for philological reasons and to clarify thoughts Mitchill proposed to strike out *azote* from the nomenclature of the day and take *septon* in its place; he also wished to expunge hydrogene and substitute phlogiston. He admitted that Priestley's experiments on zinc were difficult to explain by the antiphlogistic doctrine, adding—

It would give me great satisfaction that we could settle the points of variance on this subject; though, even as it is, I am flattered by your (Priestley's) allowing my attempt 'to reconcile the two theories to be ingenious, plausible and well-meant.... Your idea of carrying on a philosophical discussion in an amicable manner is charming'....

But the peace-maker was handling a delicate problem. He recognized this, but desired that the pioneer studies, then in progress might escape harsh polemics. This was difficult of realization for less than a month later fuel was added to the fire by Maclean, when in writing Mitchill, who had sent him Priestley's printed letter, he emphatically declared that [Pg 85]

The experiment with the zinc does not seem to be of more consequence than that with the iron and admits of an easy explanation on antiphlogistic principles.

And he further insisted that the experiments of Priestley proved water to be composed "of hydrogene and oxygen."

Four days later (July 20, 1798) Priestley wrote Mitchill that he had replaced zinc by red precipitate and did not get water on decomposing inflammable air with the precipitate. Again, August 23, 1798, he related to Mitchill

that the modern doctrine of water consisting of *oxygene* and *hydrogene* is not well founded ... water is the basis of all kinds of air, and without it no kind of air can be produced ... not withstanding the great use that the French chemists make of scales and weights, they do not pretend to weigh either their *calorique* or *light*; and why may not *phlogiston* escape their researches, when they employ the same instruments in that investigation?

There were in all eight letters sent by Priestley to Mitchill. They continued until February, [Pg 86] 1799. Their one subject was phlogiston and its rôle in very simple chemical operations. The observations were the consequence

of original and recent experiments, to which I have given a good part of the leisure of the last summer; and I do not propose to do more on the subject till I hear from the great authors of the theory that I combat in America;

but adds,—

I am glad ... to find several advocates of the system in this country, and some of them, I am confident, will do themselves honour by their candour, as well as by their ability.

This very probably was said as a consequence of the spirited reply James Woodhouse [6] made to the papers of Maclean. As known, Woodhouse worked unceasingly to overthrow the doctrine of phlogiston, but was evidently irritated by Maclean, whom he reminds—

You are not yet, Doctor, the conqueror of this veteran in Philosophy.

[Pg 87]

This was a singularly magnanimous speech on Woodhouse's part, for he had been hurling sledgehammer blows without rest at the structure Priestley thought he had reared about phlogiston and which, he believed, most unassailable, so when in 1799 (July) Priestley began his reply to his "Antiphlogistian opponents" he took occasion to remark:

I am happy to find in Dr. Woodhouse one who is equally ingenious and candid; so that I do not think the cause he has undertaken will soon find a more able champion, and I do not regret the absence of M. Berthollet in Egypt.

Noble words these for his young adversary who, in consequence of strenuous laboratory work, had acquired a deep respect and admiration for Priestley's achievements, though he considered he had gone far astray.

The various new, confirmatory ideas put forth by Priestley need not be here enumerated. They served their day.

Dr. Mitchill evidently enjoyed this controversial chemical material, for he wrote that he hoped the readers of the *Medical Repository*, in which the several papers appeared, would [Pg 88]

participate the pleasure we feel on taking a retrospect of our pages, and finding the United States the theatre of so much scientific disquisition.

And yet, when in 1800, a pamphlet of 90 pages bearing the title "The Doctrine of Phlogiston established, etc." appeared there was consternation in the ranks of American chemists. Woodhouse was aroused. He absolutely refuted every point in it experimentally, and Dr. Mitchill avowed—

We decline entering into a minute examination of his experiments, as few of his recitals of them are free from the *triune* mystery of phlogiston, which exceeds the utmost stretch of our faith; for according to it, *carbon is phlogiston*, and *hydrogen is phlogiston*, and *azote is phlogiston*; and yet there are not *three* phlogistons, but *one* phlogiston!

It was imperative to submit the preceding paragraphs on chemical topics, notwithstanding they have, in a manner, interrupted the chronological arrangement of the activities of the Doctor in his home life. They were, it is true, a part of that life—a part that every chemist will note with [Pg 89] interest and pleasure. They mean that he was not indifferent to chemistry, and that it is not to be supposed that he ever could be, especially as his visits to Philadelphia brought to his attention problems which he would never suffer to go unanswered or unsolved because of his interest in so many other things quite foreign to them. However, a backward look may be taken before resuming the story of his experimental studies.

It has already been said that the non-appearance of letters caused him anxiety. For instance he wrote Lindsey, July 28, 1796—

It is now four months since I have received any letter from you, and it gives me most serious concern.

But finally the longed-for epistle arrived and he became content, rejoicing in being able to return the news—

I do not know that I have more satisfaction from anything I ever did, than from the lay Unitarian congregation I have been the means of establishing in Philadelphia.

For the use of this group of worshipers he had engaged the Common Hall in the College (University of Pennsylvania). [Pg 90]

But amidst this unceasing activity of body and mind—very evidently extremely happy in his surroundings—he was again crushed to earth by the death of his noble wife—

Always caring for others and never for herself.

This occurred nine months after the departure of Harry. It was a fearful blow. For more than thirty-four years they had lived most happily together. The following tribute, full of deep feeling and esteem attests this—

My wife being a woman of an excellent understanding much improved by reading, of great fortitude and strength of mind, and of a temper in the highest degree affectionate and generous.... Also excelling in everything relating to household affairs, she entirely relieved me of all concern of that kind, which allowed me to give all my time to the prosecution of my studies.

She was not only a true helpmate—courageous and devoted—but certainly most desirous that the husband in whom she absolutely believed should have nothing to interrupt or arrest the pur [Pg 91] suits dear to him and in which she herself must have taken great but quiet pride, for she was extremely intelligent and original. Madam Belloc has mentioned

51

It is a tradition in the family that Mrs. Priestley once sent her famous husband to market with a large basket and that he so acquitted himself that she never sent him again!

The new house, partly planned by her, at the moment well advanced and to her fancy, was not to be her home for which she had fondly dreamed.

Priestley was deeply depressed but his habitual submission carried him through, although all this is pathetically concealed in his letters.

There were rumours flitting about that Priestley purposed returning to England. That his friends might be apprised of his real intentions the following letter was permitted to find its way into the newspapers:

Northumberland Oct. 4,
1796

My dear Sir,

Every account I have from England makes me think myself happy in [Pg 92] this peaceful retirement, where I enjoy almost everything I can wish in this life, and where I hope to close it, though I find it is reported, both here and in England that I am about to return. The two heavy afflictions I have met with here, in the death of a son, and of my wife, rather serve to attract me to the place. Though dead and buried, I would not willingly leave them, and hope to rest with them, when the sovereign disposer of all things shall put a period to my present labours and pursuits.

The advantages we enjoy in this country are indeed very great. Here we have no poor; we never see a beggar, nor is there a family in want. We have no church establishment, and hardly any taxes. This particular State pays all its officers from a treasure in the public funds. There are very few crimes committed and we travel without the least apprehension of danger. The press is perfectly free, and I hope we shall always keep out of war.

I do not think there ever was any country in a state of such rapid improvement as this at present; but we have not the same advantages for literary and philosophical pursuits that you have [Pg

93] in Europe, though even in this respect we are every day getting better. Many books are now printed here, but what scholars chiefly want are old books, and these are not to be had. We hope, however, that the troubles of Europe will be the cause of sending us some libraries and they say that it is an ill wind that blows no profit.

I sincerely wish, however, that your troubles were at an end, and from our last accounts we think there must be a peace, at least from the impossibility of carrying on the war.

With every good wish to my country and to yourself, I am, dear sir,

Yours sincerely,
J. PRIESTLEY.

Gradually the news went forth that the Doctor contemplated a second visit to the metropolis—Philadelphia, the Capital of the young Republic. He wrote—

Having now one tie, and that a strong one, to this place (Northumberland) less than I have had I propose to spend more time in Philadelphia.

[Pg 94]

As long as he was capable of public speaking it was his desire to carry forward his missionary work,

but the loss of my fore teeth (having now only two in the upper jaw) together with my tendency to stammering, which troubles me sometimes, is much against me.

Accordingly in early January of 1797 he might have been found there. He alludes in his correspondence to the presence in the city of C. Volney, a French philosopher and historian, who had been imprisoned but regained liberty on the overthrow of Robespierre when he became professor of history in the *Ecole Normal*. Volney was not particularly pleased with Priestley's discourses, and took occasion some weeks later to issue

VOLNEY'S ANSWER TO PRIESTLEY

which was advertised by the *Aurora* as on sale by the principal booksellers, price 6 cents.

He was exceedingly rejoiced at the flourishing state of the Unitarian Society and the manner in which its services were conducted.

On the occasion of his first discourse the English Ambassador, Mr. Lister, was in the audience and Priestley dined with him the day following. [Pg 95]

Friends had prevailed upon Priestley to preach a charity sermon on his next Sunday, in one of the Episcopal churches, but in the end it was "delivered at the University Hall."

His mind was much occupied with plans for controverting infidelity,

the progress of which here is independent of all reasoning, —

so he published the third edition of his "Observations on the Increase of Infidelity" and an "Outline of the Evidences of Revealed Religion." In the first of them he issued a challenge to Volney who was

much looked up to by unbelievers here.

Volney's only reply was that he would not read the pamphlet. It was in these days that Priestley saw a great deal of Thomas Jefferson; indeed, the latter attended several of his sermons. The intercourse of these friends was extremely valuable to both. Jefferson welcomed everything which Priestley did in science and consulted him much on problems of education.

At the election in the American Philosophical Society in the closing days of 1796 there was openly discussed [Pg 96]

whether to choose me (Priestley) or Mr. Jefferson, President of the Society, —

which prompted the Doctor

to give his informant good reasons why they should not choose *me*.

Naturally he listened to the political talk. He worried over the apparent dislike observed generally to France. He remarked

The rich not only wish for alliance offensive and defensive with England ... but would have little objection to the former dependence upon it,

and

The disposition of the lower orders of the people ... for the French ... is not extinguished.

He was much annoyed by Peter Porcupine. The latter was publishing a daily paper (1799) and in it frequently brought forward Priestley's name in the most opprobrious manner, although Priestley in his own words [Pg 97] —

had nothing to do with the politics of the country.

The Doctor advised friend Lindsey that

He (Porcupine) every day, advertizes his pamphlet against me, and after my name adds, "commonly known by the name of the fire-brand philosopher."

However, he flattered himself that he would soon be back in Northumberland, where he would be usefully engaged, as

I have cut myself out work for a year at least ... besides attending to my experiments.

Mr. Adams had come into the Presidency, so Priestley very properly went to pay his respects and

take leave of the late President (Washington)

whom he thought in not very good spirits, although

he invited me to Mount Vernon and said he thought he should hardly go from home twenty miles as long as he lived.

[Pg 98]

Priestley's fame was rapidly spreading through the land. Thoughtful men were doing him honor in many sections of the country, as is evident from the following clipping from a Portland (Me.) paper for March 27, 1797: —

On Friday the twenty-fourth a number of gentlemen, entertaining a high sense of the character, abilities and services of the Reverend Doctor JOSEPH PRIESTLEY, as a friend and promoter of true science dined together at the Columbian Tavern, in commemoration of his birth. The following toasts were given.

1. That Illustrious Christian and Philosopher, Joseph Priestley: May the world be as grateful to him for his services as his services are beneficial to the world.

2. May the names of Locke, Newton, Montesquieu, Hartley and Franklin be had in everlasting remembrance.

3. The great gift of God to man, Reason! May it influence the world in policy, in laws, and in religion.

4. TRUTH: May the splendour of her charms dissipate the gloom of superstition, and expel hypocricy from the heart of man. [Pg 99]

5. May our laws be supported by religion: but may religion never be supported by law.

6. White-robed Charity: May she accompany us in all our steps and cover us with a mantle of love.

7. Christians of all denominations: May they "love one another."

As it was a "feast of reason" the purest philanthrophy dignified the conversation; and moderation and temperance bounded every effusion of the heart.

It was in the summer of 1797 that he carried forward his work on Phlogiston, alluded to on p. 81. He understood quite well that the entire chemical world was against him but he was not able to find good reasons

to despair of the old system.

It must be remembered that in these days, also, he had Thomas Cooper with him. With this gentleman he discussed his scientific studies and with him also he carried on many arguments upon the burning subject of infidelity, about which he continuously wrote his friends in this country and in England. It was quite generally believed that [Pg 100] Cooper was an infidel. Never, however, did their intimacy suffer in the slightest by their conflicting views.

The *Church History* continued to hold Priestley's first thought. He was a busy student, occupied with a diversity of interests and usually cheerful and eager to follow up new lines of endeavor. The arrival of vessels from the home country was closely watched. Books and apparatus were brought by them. While, as observed, he was singularly cheerful and happy, he confessed at times that

> my character as a philosopher is under a cloud.

Yet, this was but a momentary depression, for he uttered in almost the same breath —

> Everything will be cleared up in a reasonable time.

Amid the constant daily duties he found real solace in his scientific pursuits; indeed when he was quite prepared to abandon all his activities he declared of his experiments that he could not stop them for [Pg 101]

> I consider them as that study of the works of the great Creator, which I shall resume with more advantage hereafter.

He advised his friends Lindsey and Belsham —

> I cannot express what *I* feel on receiving your letters. They set my thoughts afloat, so that I can do nothing but ruminate a long time; but it is a most pleasing melancholy.

Far removed from European events he was nevertheless ever keen and alert concerning them. Then the winter of 1797 appears to have been very severe. His enforced confinement to home probably gave rise to an introspection, and a slight disappointment in matters which had formerly given him pleasure. For example, he puzzled over the fact that on his second visit to Philadelphia, Mr. Adams was present but once at his lectures, and remarks —

> When my lectures were less popular, and he was near his presidentship, he left me, making a kind of apology, from the members of the principal Presbyterian Church having offered him a pew there. He seem [Pg 102] ed to interest himself in my favour against M. Volney, but did not subscribe to my Church History ... I suppose he was not pleased that I did not adopt his dislike of the French.

When January of 1798 arrived his joy was great. A box of books had come. Among them was a General Dictionary which he regarded as a real treasure. Reading was now his principal occupation. He found the making of many experiments irksome and seemed, all at once, "quite averse to having his hands so much in water." Presumably these were innocent excuses for his devotion to the Church History which had been brought up to date. Furthermore he was actually contemplating transplanting himself to France. But with it all he wrote assiduously on religious topics, and was highly pleased with the experimental work he had sent to Dr. Mitchill (p. 85).

He advised his friends of the "intercepted letters" which did him much harm when they were published. They called down upon him severest judgement and suspicion, and made him —

disliked by all the friends of the ruling power in this country.

[Pg 103]

It may be well to note that these "intercepted letters" were found on a Danish ship, inclosed in a cover addressed to

DR. PRIESTLEY, IN AMERICA

They came from friends, English and French, living in Paris. They abounded

with matter of the most serious reflection.... If the animosity of these apostate Englishmen against their own country, their conviction that no submissions will avert our danger, and their description of the engines employed by the Directory for our destruction, were impressed as they ought to be, upon the minds of all our countrymen, we should certainly never again be told of the innocent designs of these traitors, or their associates —

The preceding quotation is from a booklet containing exact copies of the "intercepted letters."

In the first of the letters, dated Feb. 12, 1798, the correspondent of Priestley tells that he had met a young Frenchman who had visited Northumberland [Pg 104]

and we all rejoiced at the aggreeable information that at the peace you would not fail to revisit Europe; and that he hoped you would fix yourself in this country (France). Whether you fix yourself here or in England, (*as England will then be*) is probably a matter of little importance ... but we all think you are misplaced where you are, though, no doubt, in the way of *usefulness* —

The editor of the letters annotates *usefulness* thus:

Dr. Priestley is *in the way of usefulness* in America, because he is labouring there, as his associates are in Europe, to disunite the people from their government, and to introduce the blessings of French anarchy.

These "intercepted letters" in no way prove that Dr. Priestley was engaged in any movement against his native land or against his adopted country. However, the whole world was in an uproar. People were ready to believe the worst regarding their fellows, so it is not surprising that he should have declared himself "disliked."

He alludes frequently to the marvelous changes [Pg 105] taking place in the States. Everything was in rapid motion. Taxes were the topic of conversation on all sides.

To divert his philosophizing he busied himself in his laboratory where many "original experiments were made." He avoided the crowd. There was too great a party spirit. Indeed, there was violence, so he determined not to visit Philadelphia. He sought to escape the "rancorous abuse" which was being hurled at him —

as a citizen of France.

One must read his correspondence to fully appreciate Priestley during the early days of 1799. What must have been his mental condition when he wrote Lindsey —

As to a public violent death the idea of that does not affect me near so much

and

I cannot express what I feel when I receive and read your letters. I generally shed many tears over them.

There was no assurance in financial and commercial circles. The hopes of neither the more [Pg 106] sober, nor of the wild and fanatic reformers of humanity could be realized, and they got into such a war of hate and abuse that they themselves stamped their doctrines false.

Priestley was out of patience with the public measures of the country. He disliked them as much as he did those of England, but added

Here the excellence of the Constitution provides a remedy, if the people will make use of it, and if not, they deserve what they suffer.

The Constitution was a favorite instrument with him. A most interesting lecture upon it will be found among the *Discourses* which he proposed delivering in Philadelphia. This never occurred.

The Academy he expected to see in operation failed for support. The walls were raised and he feared it would go no further. The Legislature had voted it $3000, but the Senate negatived this act. He thought of giving up the presidency of it.

He wrote Dr. Rush that he was quite busy with replies to Dr. Woodhouse's attack on his confirmation of the existence of phlogiston, (p. 88). He relished his discussions with Woodhouse and was confident that eventually he would "overturn the [Pg 107] French system of chemistry." He further remarked to Rush—

Were you at liberty to make an excursion as far as these *back woods* I shall be happy to see you, and so would many others.

But at that particular moment Rush was too much engaged in combating yellow fever, which again ravaged Philadelphia, and all who could, fled, and the streets were "lifeless and dead." The prevalence of this fearful plague was a potent factor in Priestley's failure to visit the City during the year—the last year of a closing Century which did not end in the prosperity anticipated for it in the hopeful months and years following the war. It seemed, in many ways, to be the end of an era. Washington died December 14, 1799, and the Federalists' tenure of power was coming to a close. The Jeffersonians, aided by eight of the electoral votes of Pennsylvania, won the victory, amid outbursts of unprecedented political bitterness. It was, therefore, very wise that the Doctor remained quietly at home in Northumberland with his experiments and Church History.

The new Century—the 19th—found our beloved philosopher at times quite proud of the success he had with his experiments and full of [Pg 108] genuine hope that "phlogiston" was established; and again dejected because of the "coarse and low articles" directed against him by the prints of the day. To offset, in a measure, the distrust entertained for him because of the "intercepted letters" he addressed a series of *Letters* to the inhabitants of Northumberland

and vicinity. These were explanatory of his views. At home they were most satisfying but in the city they brought upon him "more abuse." And, so, he translated a passage from Petrarch which read —

By civil fueds exiled my native home,
Resign'd, though injured, hither I have come.
Here, groves and streams, delights of rural ease;
Yet, where the associates, wont to serve and please;
The aspect bland, that bade the heart confide?
Absent from these, e'en here, no joys abide.

And these were incorporated in his brochure.

Having alluded to the *Letters* addressed to the Northumberland folks, it may be proper to introduce a letter which Priestley received from Mr. Jefferson, whom the former was disposed to hold [Pg 109] as "in many respects the first man in this Country:"

Philadelphia, Jan. 18, 1800.
Dear Sir —

I thank you for the pamphlets (Letters) you were so kind as to send me. You will know what I thought of them by my having before sent a dozen sets to Virginia, to distribute among my friends; yet I thank you not the less for these, which I value the more as they came from yourself.

The papers of Political Arithmetic, both in yours and Mr. Cooper's pamphlets, are the most precious gifts that can be made to us; for we are running navigation-mad, and commerce-mad, and Navy-mad, which is worst of all. How desirable it is that you should pursue that subject for us. From the porcupines of our country you will receive no thanks, but the great mass of our nation will edify, and thank you.

How deeply have I been chagrined and mortified at the persecutions which fanaticism and monarchy have excited against you, even here! At first, I believed it was merely a continuance of the English persecution; but I observe that, on the demise [Pg 110] of Porcupine, and the division of his inheritance between Fenno and Brown, the latter (though succeeding only to the Federal portion of Porcupinism, not the Anglican, which is Fenno's part) serves up for the palate of his sect dishes of abuse against you as high-seasoned as Porcupine's were. You have sinned against Church and King, and therefore can never be forgiven. How sincerely I have regretted that your friend, before he fixed a choice of position, did not visit the valleys on each side of the blue range in Virginia, as Mr. Madison and myself so much wished. You would have found there equal soil, the finest climate, and the most healthy air on the earth, the homage of universal reverence and love, and the power of the country spread over you as a shield; but, since you would not make it your Country by adoption, you must now do it by your good offices.

Mr. Livingston, the Chancellor of New York, so approved the "Letters" that he got a new edition of them printed at Albany.

The following letter to this same gentleman, although upon another subject than the "Letters" [Pg 111] is not devoid of interest. It has come into the writer's hands through the kind offices of Dr. Thomas L. Montgomery, State Librarian of Pennsylvania:

Sir,

I think myself much honoured by your letter, and should have thought myself singularly happy if my situation had been near to such a person as you. Persons engaged in scientific pursuits are few in this country. Indeed, they are not very numerous anywhere. In other respects I think myself very happy where I am.

I have never given much attention to machines of any kind, and therefore cannot pretend to decide concerning your proposal for the improvement of the fire engine. It appears to me to deserve attention. But I do not for want of a drawing see in what manner the steam is to be let into the cylinder, or discharged from it. There would be, I fear, an objection to it from the force necessary to raise the column of mercury, and from the evaporation of the mercury in the requisite heat. I have found that it loses weight in 70° Fahrenheit. If the mercury was pure, I should not [Pg 112] apprehend much from the calcination of it, though, as I have observed, the agitation of it in water, converts a part of it into a black powder, which I propose to examine farther.

If travelling was attended with no fewer inconveniences here than it is in England, I should certainly wait upon you and some other friends at New York. But this, and my age, render it impossible, and it would be unreasonable to expect many visitors in this *back woods*.

I shall be very happy to be favoured with your correspondence, and am,

Sir,
Yours sincerely,
J. PRIESTLEY
Northumberland April 16, 1799.

In this period Thomas Cooper was convicted of libel. He was thrown into prison. Priestley regarded him as a rising man in the Country. [7] He said the act was the last blow of the Federal party "which is now broke up." [Pg 113]

Priestley's daughter, in England, was ill at this time. Her life was despaired of and tidings from her were few and most distressing, but the Doctor maintained a quiet and calm assurance of her recovery.

Subsequent correspondence between Mr. Jefferson and Priestley had much in it about the new College which the former contemplated for the State of Virginia. Indeed, the thought was entertained that Priestley himself might become a professor in it, but his advanced age, he contended forbade this, although he was agreeable to the idea of getting professors from Europe.

Here, perhaps, may well be included several letters, now in possession of the Library of Congress, which reveal the attitude of Dr. Priestley toward President Jefferson, who was indeed most friendly to him:

Dear Sir —

I am flattered by your thinking so favourably of my *pamphlets*, which were only calculated to give some satisfaction to my suspicious neighbours. Chancellor Livingston informs me that he has got an edition of them printed at Albany, for the information of the people in the back country, where, he says, it is so much wanted. Indeed, it [Pg 114] seems extraordinary, that in such a country as this, where there is no court to dazzle men's eyes a maxim as plain as that 2 and 2 make 4 should not be understood, and acted upon. It is evident that the bulk of mankind are governed by something very different from reasoning and argument. This principle must have its influence even in your Congress, for if the members are not convinced by the excellent speeches of Mr. Gallatin and Nicolas, neither would they be persuaded tho one should rise from the dead.

It is true that I had more to do with colleges, and places of education, than most men in Europe; but I would not pretend to advise in this country. I will, however, at my leisure, propose such *hints* as shall occur to me; and if you want tutors from England, I can recommend some very good ones. Were I a few years younger, and more moveable, I should make interest for some appointment in your institution myself; but age and inactivity are fast approaching, and I am so fixed here, that a remove is absolutely impossible, unless you were possessed of _Aladin_'s *lamp*, and could trans [Pg 115] port my house, library, and laboratory, into Virginia without trouble or expense.

On my settlement here the gentlemen in the neighbourhood, thinking to make me of some use, set on foot a college, of which I gave them the plan, and they got it incorporated, and made me the president; but tho I proposed to give lectures *gratis*, and had the disposal of a valuable library at the decease of a learned friend (new, near so), and had it in my power to render them important service in various ways, yet, owing I suspect, in part at least, to religious and political prejudices, nothing more has been done, besides marking the site of a building these five years, so that I have told them I shall resign.

I much wish to have some conversation with you on social subjects; but I cannot expect that the Vice President of the United States should visit me in my *shed* at Northumberland, and I cannot come to you. I intended on my settling here to have spent a month or so every winter at Philadelphia, but the state of the times, and various accidents, have a little deranged my finances, and I prefer to spend what I can spare on [Pg 116] my experiments, and publication, rather than in travelling and seeing my friends.

With the greatest respect, I am,
Dear Sir,

Yours sincerely,
J. PRIESTLEY.

Northumberland Jan. 30, 1800.

Dear Sir—

I enclose my thoughts on the subject you did me the honour to propose to me. Your own better judgment will decide concerning their value, or their fitness for the circumstances of your College. This may require a very different distribution of the business from that which I here recommend.

I thank you for your care to transmit a copy of my works to Bp. Madison. He, as well as many others, speaks of the increasing spread of republican principles in this country. I wish I could see the effects of it. But I fear we flatter ourselves, and if I be rightly informed, my poor *Letters* have done more harm than good. I can only say that I am a sincere well wisher to this country, and the purity and stability of its constitution.

Yours sincerely,
J. PRIESTLEY.

Northumberland May 8, 1800. [Pg 117]

Hints Concerning Public Education

Persons educated at public seminaries are of two classes. One is that of professional men, and physicians and divines who are to be qualified for entering upon their professions immediately after leaving the college or university. The other is that of gentlemen, and those who are designed for offices of civil and active life. The former must be minutely instructed in everything adding to their several professions, whereas to the latter a general knowledge of the several branches of science is sufficient. To the former, especially that of Medicine, several professors are necessary, as the business must be subdivided, in order to be taught to advantage. For the purpose of the latter fewer professors are wanted, as it is most advisable to give them only the elements of the several branches of knowledge, to which they may afterwards give more particular attention, as they may have a disposition or convenience for it.

Lawyers are not supposed to be qualified for entering upon their professions at any place of public education. They are therefore to be considered as gentlemen to whom a general knowledge is sufficient. It is advisable, however, that when any subject, as that of Medicine, is much divided, [Pg 118] and distributed among a number of professors, lectures of a more general and popular nature be provided for the other classes of students, to whom some knowledge of the subject may be very useful. A general knowledge, for example, of anatomy and of medicine, too, is useful to all persons, and therefore ought not to be omitted in any scheme of liberal education. And if in a regular school of medicine any of the professors would undertake this, it might serve as an useful introduction to that more particular and accurate knowledge which is necessary for practiced physicians.

The branches of knowledge which are necessary to the teachers of religion are not so many, or so distinct from each other, but that they may all be taught by one professor, as far as is necessary to qualify persons for commencing preachers. To acquire more knowledge, as that of the scriptures, ecclesistical history, etc. must be the business of their future lives. But every person liberally educated should have a general knowledge of Metaphysics, the theory of morals, and religion; and therefore some popular lectures of this kind should be provided for the students in general.

One professor of antient languages may be sufficient for a place of liberal education, and I would not make any provision for instruction in the [Pg 119] modern languages, for tho the knowledge of them, as well as skill in fencing, dancing and riding, is proper for gentlemen liberally educated, instruction in them may be procured on reasonable terms without burdening the funds of the seminary with them.

Abstract Mathematics, and Natural Philosophy, are so distinct, that they require different teachers. One is sufficient for the former, but the latter must be subdivided, one for natural history, another for experimental Philosophy in general and a third for chemistry; in consequence of the great extension of this branch of experimental Philosophy of late years. The botany, mineralogy, and other branches of natural history are sufficiently distinct to admit of different professors, nothing more than a general knowledge of each of them, and directions for acquiring a more extended knowledge of them is necessary at any place of education.

Two or three Schools of Medicine I should think sufficient for all the United States for some years to come, but with respect to these I do not pretend to give any opinion not having sufficient knowledge of the subject. Places of liberal education in general should be made more numerous, and for each of them I should think the following professors (if the funds of the Society will admit of it) should be engaged, *viz.* (1) For the antient [Pg 120] languages. (2) The Belles Lettres, including universal Grammar, Oratory, criticism and bibliography. (3) Mathematics. (4) Natural history. (5) Experimental Philosophy. (6) Chemistry, including the theory of Agriculture. (7) Anatomy and Medicine. (8) Geography and history, Law, and general policy. (9) Metaphysics, morals, and theology.

A course of liberal education should be as comprehensive as possible. For this purpose a large and well chosen *library* will be of great use. Not that the students should be encouraged to read books while they are under tuition, but an opportunity of seeing books, and looking into them, will give them a better idea of the value of them than they could get by merely hearing of them, and they would afterwards better know what books to purchase when they should have the means and the leisure for the perusal of them. A large collection of books will also be useful to the lecturer in *bibliography* and would recommend the seminary to the professors in general, and make it a desirable place of residence for gentlemen of a studious turn.

2. In order to engage able professors, some fixed salaries are necessary; but they should not be much more than a bare subsistence. They will then have a motive to exert themselves, and by [Pg 121] the fees of students their emoluments may be ample. The professorships in the English universities, which are largely endowed, are sinecures; while those in Scotland, to which small stipends are annexed, are filled by able and active men.

3. It is not wise to engage any persons who are much advanced in life, or of established reputation for efficient teachers. They will not be so active as younger men who have a character to acquire. They

will also better accommodate their lectures to the increasing light of the age, whereas old men will be attached to old systems, tho ever so imperfect. Besides, they are the most expert in teaching who have lately learned, and the minutae of science, which are necessary to a teacher, are generally forgotten by good scholars who are advanced in life, and it is peculiarly irksome to relearn them.

4. I would not without necessity have recourse to any foreign country for professors. They will expect too much deference, and the natives will be jealous of them.

5. Three things must be attended to in the education of youth. They must be *taught*, *fed* and *governed* and each of these requires very different qualifications. They who are the best qualified to teach are often the most unfit to govern, and it is generally advisable that neither of these have [Pg 122] anything to do with providing victuals. In the English universities all these affairs are perfectly distinct. The *tutors* only teach, the *proctors* superintend the discipline, and the *cooks* provide the victuals.

Philadelphia, Apr. 10, 1801.

Dear Sir—

Your kind letter, which, considering the numerous engagements incident to your situation, I had no right to expect, was highly gratifying to me, and I take the first opportunity of acknowledging it. For tho I believe I am completely recovered from my late illness, I am advised to write as little as possible. Your invitation to pay you a visit is flattering to me in the highest degree, and I shall not wholly despair of some time or other availing myself of it, but for the present I must take the nearest way home.

Your resentment of the treatment I have met with in this country is truly generous, but I must have been but little impressed with the principles of the religion you so justly commend, if they had not enabled me to bear much more than I have yet suffered. [Pg 123] Do not suppose that, after the much worse treatment to which I was for many years exposed in England (of which the pamphlet I take the liberty to inclose will give you some idea) I was much affected by this. My *Letters to the Inhabitants of Northumberland* were not occasioned by any such thing, tho it served me as a pretense for writing them, but the threatenings of Mr. Pickering, whose purpose to send me out of the country Mr. Adams (as I conclude from a circuitous attempt that he made to prevent it) would not, in the circumstances in which he then was, have been able to directly oppose. My publication was of service to me in that and other respects and I hope, in some measure, to the common cause. But had it not been for the extreme absurdity and violence of the late administration, I do not know how far the measures might not have been carried. I rejoice more than I can express in the glorious reverse that has taken place, and which has secured your election. This I flatter myself will be the permanent establishment of truly republican principles in this country, and also contribute to the same desirable event in more distant ones. [Pg 124]

I beg you would not trouble yourself with any answer to this. The knowledge of your good opinion and good wishes, is quite sufficient for me. I feel for the difficulties of your situation, but your spirit and prudence will carry you thro them, tho not without paying the tax which the wise laws of nature have imposed upon preeminence and celebrity of every kind, a tax which, for want of true greatness of mind, neither of your predecessors, if I estimate their characters aright, paid without much reluctance.

With every good wish, I am,
Dear Sir,
Yours sincerely,
J. PRIESTLEY.

P.S.

As I trust that *Politics* will not make you forget what is due to *science*, I shall send you a copy of some articles that are just printed for the *Transactions of the Philosophical Society* in this place. No. (5) p. 36 is the most deserving of your notice. I should have sent you my *Defence of Phlogiston*, but that I presume you have seen it.

[Pg 125]

June, 1802.
To Thomas Jefferson, President of the United States of America.

Sir,

My high respect for your character, as a politician, and a man, makes me desirous of connecting my name, in some measure with yours while it is in my power, by means of some publication, to do it.

The first part of this work, which brought the history to the fall of the western empire, was dedicated to a zealous friend of civil and religious liberty, but in a private station. What he, or any other friend of liberty in Europe, could only do by their good wishes, by writing, or by patriot suffering, you, Sir, are actually accomplishing, and upon a theatre of great and growing extent.

It is the boast of this country to have a constitution the most favourable to political liberty, and private happiness, of any in the world, and all say that it was yourself, more than any other individual, that planned and established it; and to this opinion your conduct in various public offices, and now in the highest, gives the clearest attestation. [Pg 126]

Many have appeared the friends of the rights of man while they were subject to the power of others, and especially when they were sufferers by it; but I do not recollect one besides yourself who retained the same principles, and acted by them, in a station of real power. You, Sir, have done more than this; having proposed to relinquish some part of the power which the constitution gave you; and instead of adding to the burden of the people, it has been your endeavour to lighten those burdens tho the necessary consequence must be the diminution of your influence. May this great example, which I doubt not will demonstrate the practicability of truly republican principles, by the actual existence of a form of government calculated to answer all the useful purposes of government (giving equal protection to all, and leaving every man in the possession of every power that he can exercise to his own advantage, without infringing on the equal liberty of others) be followed in other countries, and at length become universal.

Another reason why I wish to prefix your name to this work, and more appropriate to the subject of it, is that you have ever [Pg 127] been a strenuous and uniform advocate of religious no less than civil liberty, both in your own state of Virginia, and in the United States in general, seeing in the clearest light the various and great mischiefs that have arisen from any particular form of religion being favoured by the State more than any other; so that the profession or practice of religion is here as free as that of philosophy, or medicine. And now the experience of more than twenty years leaves little room to doubt but that it is a state, of things the most favourable to mutual candour, which is of great importance to domestic peace and good neighbourhood and to the cause of all truth, religious truth least of all excepted. When every question is thus left to free discussion, there cannot be a doubt but that truth will finally prevail, and establish itself by its own evidence; and he must know little of mankind, or of human nature, who can imagine that truth of any kind will be ultimately unfavourable to general happiness. That man must entertain a secret suspicion of his own principles who wishes for any exclusive advantage in his defence or profession of them. [Pg 128]

Having fled from a state of persecution in England, and having been exposed to some degree of danger in the late administration here, I naturally feel the greater satisfaction in the prospect of passing the remainder of an active life (when I naturally wish for repose) under your protection. Tho arrived at the usual term of human life it is now only that I can say I see nothing to fear from the hand of power, the government under which I live being for the first time truly favourable to me. And tho it will be evident to all who know me that I have never been swayed by the mean principle of fear, it is certainly a happiness to be out of the possibility of its influence, and to end ones days in peace, enjoying some degree of rest before the state of more perfect rest in the grave, and with the hope of rising to a state of greater activity, security and happiness beyond it. This is all that any man can wish for, or have; and this, Sir, under your administration, I enjoy.

With the most perfect attachment, and every good wish I subscribe myself not your subject, or humble servant, but your sincere admirer.

J. PRIESTLEY.

[Pg 129]

Dear Sir,

As there are some particulars in a letter I have lately received from Mr. Stone at Paris which I think it will give you pleasure to have, and Mr. Cooper has been so obliging as to translate them for me, I take the liberty to send them, along with a copy of my *Dedication*, with the correction that you suggested, and a Note from the latter with which you favoured me concerning what you did with respect to the *constitution*, and which is really more than I had ascribed to you. For almost everything of importance to political liberty in that instrument was, as it appears to me, suggested by you, and as this was unknown to myself, and I believe is so with the world in general, I was unwilling to omit this opportunity of noticing it.

I shall be glad if you will be so good as to engage any person sufficiently qualified to draw up such an account of the *constitutional forms* of this country as my friends say will be agreeable to the emperor, and I will transmit it to Mr. Stone.

Not knowing any certain method of sending a letter to France and presuming that you do I take the liberty to inclose my letter [Pg 130] to Mr. Stone. It is, however, so written that no danger can arise to him from it, into whatever hands it may fall.

The state of my health, though, I thank God, much improved, will not permit me to avail myself of your kind invitation to pay you a visit. Where ever I am, you may depend upon my warmest attachment and best wishes.

J. PRIESTLEY.

Northumberland Oct. 29, 1802.

P.S.

I send a copy of the *Preface* as well as of the *Dedication*, that you may form some idea of the work you are pleased to patronize.

Northumberland Jan. 25, 1803.

Dear Sir,

As you were pleased to think favourably of my pamphlet entitled *Socrates and Jesus compared*, I take the liberty to send you a *defence* of it. My principal object, you will perceive, was to lay hold of the opportunity, given me by Mr. B. Linn, to excite some attention to doctrines which I consider as of peculiar importance in the [Pg 131] Christian system, and which I do not find to have been discussed in this country.

The Church History is, I hope, by this time in the hands of the bookseller at Philadelphia, so that you will soon, if my directions have been attended to, receive a copy of the work which I have the honour to dedicate to you.

With the greatest respect and attachment, I am

Dear Sir,
Yours sincerely,
J. PRIESTLEY.

Dear Sir,

I take the liberty to send you *a second defence of my pamphlet about Socrates*, on the 16th page of which you will find that I have undertaken the task you were pleased to recommend to me. On giving more attention to it, I found, as the fox did with respect to the lion, that my apprehensions entirely vanished. Indeed, I have already accomplished a considerable part of the work, and in about a year from this time I hope to finish the whole, provided my health, which is very precarious, be continued in the state in which it now is. [Pg 132] I directed a copy of the *tract on phlogiston* to be sent to you from Philadelphia, and I shall order another, which, together with the inclosed papers, I shall be much obliged to you if you will convey to. Mr. Livingston. Please also to cast an eye over them yourself; and if you can with propriety promote my interest by any representation of yours, I am confident you will do it.

When you wrote to me at the commencement of your administration, you said "the only dark speck in our horizon is in Louisiana." By your excellent conduct it is now the brightest we have to look to.

Mr. Vaughan having applied to me for a copy of my Harmony of the Evangelists, which was not to be had in Philadelphia, and intimated that it was for you, my son, whose copy is more perfect than mine, begs the honour of your acceptance of it, as a mark of his high esteem, in which he has the hearty concurrence of

Dear Sir,
Yours sincerely,
J. PRIESTLEY.

Northumberland Dec. 12, 1803.

[Pg 133]

His European correspondents were informed that he was much engaged with religious matters. While his theological views were not received very graciously yet he found

some young men of a serious and inquisitive turn, who read my works, and are confirmed Unitarians.

In one of his communications to Lindsey, written in April 1800, he expresses himself in the following most interesting way relative to his scientific engagements. American men of science will welcome it: This is the message:

I send along with this an account of a course of experiments of as much importance as almost any that I have ever made. Please to shew it to Mr. Kirwan, and give it either to Mr. Nicholson for his journal, or to Mr. Phillips for his magazine, as you please. I was never more busy or more successful in this way, when I was in England; and I am very thankful to Providence for the means and the leisure for these pursuits, which next to theological studies, interest me the most. Indeed, there is a natural alliance between them, as there must be between the word and the works of God.

[Pg 134]

He was now at work apparently in his own little laboratory adjacent to his dwelling place. For more than a century this structure has remained practically as it was in the days of Priestley. In it he did remarkable things, in his judgment; thus refuting the general idea that after his arrival in America nothing of merit in the scientific direction was accomplished by him. The satisfactory results, mentioned to Lindsey, were embodied in a series of "Six Chemical Essays" which eventually found their way into the Transactions of the American Philosophical Society. It is a miscellany of observations. In it are recorded the results found on passing the "vapour of spirit of nitre" over iron turnings, over copper, over perfect charcoal, charcoal of bones, melted lead, tin and bismuth; and there appears a note to the effect that in Papin's digester "a solution of caustic alkali, aided by heat, made a *liquor silicum* with pounded flint glass." There is also given a description of a pyrophorus obtained from iron and sulphur. More interesting, however, was the account of the change of place in different kinds of air, "through several interposing substances," in which Priestley recognized distinctly for the first time, the phenomena of gaseous diffusion. There are also references to the absorption of air by water, and of course, as one [Pg 135] would expect from the Doctor, for it never failed, there is once more emphasized "certain facts pertaining to phlogiston." His friends were quite prepared for such statements. They thought of Joseph Priestley and involuntarily there arose the idea of phlogiston.

The little workshop or laboratory, in Northumberland, where these facts were gathered, will soon be removed to the Campus of Pennsylvania State College. It will be preserved with care and in it, it is hoped, will be gradually assembled everything to be found relating to the noble soul who once disclosed Nature's secrets in this simple primitive structure, which American chemists should ever cherish, and hold as a Mecca for all who would look back to the beginnings of chemical research in our beloved country.

How appropriate it would be could there be deposited in the little laboratory, the apparatus owned and used by Priestley, which at present constitutes and for many years past has formed an attractive collection in Dickinson College, (Pa.) There would be the burning lens, the reflecting telescope, the refracting telescope (probably one of the first achromatic telescopes made), the air-gun, the orrery, and flasks with heavy ground necks, and heavy curved tubes with ground stoppers—all brought (to Dickinson) through the [Pg 136] instrumentality of Thomas Cooper, "the greatest man in America in the powers of his mind and acquired information and that without a single exception" according to Thomas Jefferson.

And how the Library would add to the glory of the place, but, alas! it has been scattered far and wide, for in 1816, Thomas Dobson advertised the same for sale in a neatly printed pamphlet of 96 pages. In it were many scarce and valuable books. The appended prices ranged quite widely, reaching in one case the goodly sum of two hundred dollars!

And as future chemists visit this unique reminder of Dr. Priestley it should be remembered that on the piazza of the dwelling house there assembled August 1, 1874, a group of men who planned then and there for the organization of the present American Chemical Society.

The "Essays," previously mentioned, will be found intensely interesting but they are somewhat difficult to read because of their strange nomenclature. Here is Priestley's account of the method pursued by him to get nitrogen:

Pure phlogisticated air (nitrogen) may be procured in the easiest and surest manner by the use of iron only—To do this I fill phials with turnings of malleable iron, [Pg 137] and having filled them with water, pour it out, to admit the air of the atmosphere, and in six or seven hours it will be diminished ... what remains of the air in the phials will be the purest phlogisticated air (nitrogen).

Among his contributions to the scientific periodicals of the times there was one relating to the sense of hearing. It is a curious story. One may properly ask whether the singular facts in it were not due to defects in Priestley's own organs of hearing. The paper did not arouse comment. It was so out of the ordinary experimental work which he was carrying forward with such genuine pleasure and intense vigour.

Strong appeals were steadily coming from English friends that he return. While commenting on the pleasure he should have in seeing them he firmly declared that the step would not be wise. In short, despite all arguments he had determined to

remain where I am for life.

The prejudices against him were abating, although he said [Pg 138]

that many things are against me; and though they do not *shake* my faith, they *try* it.

There had gathered a class of fourteen young men about him in the Northumberland home. They had adopted his Unitarian ideas.

To them he lectured regularly on theology and philosophy. Those must have been inspiring moments. It was in this wise that the aged philosopher felt he was doing good and was most useful. He said that it was

a pretty good class of young men to lecture to.

Much time was given to his English correspondents. Them he advised of the rapid development of the States. He sent to some pictures of the country about him, and with much delight he referred to the fact that Jefferson, whom he ardently admired, was now, in the closing weeks of 1800, the President, and his associate—Aaron Burr, Vice-President. He announced to English friends that the late administration, that of John Adams, was

almost universally reprobated.

Mr. Jefferson, he insisted, "will do nothing rashly,"

His being president may induce me to visit the federal city, and perhaps his seat in Virginia.

The seat of government, as may be inferred, had been removed to Washington from Philadelphia. But to the latter center, which still offered many attractions, Priestley journeyed for the third time early in 1801. He was not especially desirous of making this third visit, but as his son and daughter came down a distance of 130 miles on business, he determined to accompany them. True, Congress was no longer there, but there were many interesting people about with whom he had great pleasure. With Bishop White, who was most orthodox and whom he saw frequently, he enjoyed much "Christian and edifying conversation." John Andrews was another favorite. He was a violent Federalist and informed Priestley that the latter

had done them (the Federalists) more mischief than any other man,

yet these two noble spirits lived in amity, and Priestley several times announced that Dr. Andrews was a Unitarian, which is not the thought today in regard to the latter.

It was an eventful year—this year of 1801. Much that was unexpected happened. It brought joy and it brought sorrow.

Perhaps it would be just as well to note the scientific progress of the Doctor during this year, for he gave forth the statement that he had succeeded in producing air by freezing water. This production of air was one of his earlier ideas (p. 62), and now he wrote—

The harder the frost was the more air I procured.

Further, he announced that on heating manganese (dioxide) in inflammable air

no water is formed,

and what is rather astounding, he was certain that *azote* consisted of hydrogen and oxygen.

To the *Medical Repository*, which he regarded highly, there was sent a rather thoughtful disquisition on dreams. In it the idea was expressed

that dreams have their seat in some region of the brain more deeply seated than that which is occupied by our waking thoughts.

[Pg 141]

A "Pile of Volta" had been sent out from England. It amused him and he studied it carefully when he was led to remark upon the theory of this curious process as follows:

The operation wholly depends on the calcination of the zinc, which suffers a great diminution in weight, while the silver is little affected, and all metals lose their phlogiston in calcination, therefore what remains of the zinc in metallic form in the pile and everything connected with that end of it, is supersaturated with phlogiston.

More need not be quoted. It was phlogiston and that only which occasioned the electric current. It may properly be added that in this connection he wrote:

It is said the inventor of the galvanic pile discovered the conducting power of charcoal, whereas it was one of my first observations in electricity, made in 1766.

Some additional attention to air was also given by him, and in so doing he reached the conclusion that

The diamond and charcoal of copper are, as nearly as possible, pure phlogiston.

[Pg 142]

One wonders how he could so persuade himself, for these bodies surely possessed weight. Why did he not rely more upon his balance?

With Woodhouse he discussed the product from passing water over heated charcoal. He had been endeavoring to refute certain statements made by Cruikshank. There is no question but that he had carbon monoxide in hand, and had it as early as 1799, and that he had obtained it in several different ways. Observe this statement:

I always found that the first portion of the heavy inflammable air, resulting from the passage of steam over heated charcoal was loaded with fixed air (CO_2), but that in the course of the process this disappeared, the remaining air (CO) burning with a lambent flame.

Scarcely had Priestley set foot in Philadelphia on his third visitation than the *Port Folio*, devoted usually to literature and biography, printed the following unkind words:

The tricks of Dr. Priestley to embroil the government, and disturb the religion of his own country, have not the merit of novelty.

[Pg 143]

To which the *Aurora* replied:

When Porcupine rioted in the filth of a debauched and corrupt faction in this city, no person experienced so much of his obscene and vulgar abuse as Dr. Priestley. There is not a single fact on record or capable of being shewn, to prove that Dr. Priestley was guilty of any other crime than being a dissenter from the church of England, and a warm friend of American Independence. For this he was abused by Porcupine—and Denny is only Porcupine with a little more tinsel to cover his dirt. It is worthy of remark, that after a whole sheet of promises of "literary lore" and "products of the master of spirits" of the nation—the first and second numbers of the

Portable Foolery, are stuffed with extracts from British publications of an ordinary quality.

The attack of the Port Folio was most ungracious. It may have been due to irritation caused by the appearance of a second edition of Priestley's "Letters to the Inhabitants of Northumberland." Nevertheless the thoughtful and dignified men of the City—men who admired [Pg 144] Priestley's broad catholic spirit and brave attitude upon all debatable questions, men who appreciated his scientific attainments, invited him to the following subscription dinner, as announced in the *Aurora*, March, 6th:

At 4 o'clock in the afternoon about one hundred citizens sat down to an elegant entertainment prepared by Mr. Francis to celebrate the commencement of the administration of Mr. Jefferson. The Governor honored the company with his presence. Several respectable Foreigners were invited to partake of the festival.... A variety of patriotic songs were admirably sung; and the following toasts were drank with unanimous applause.

1. The Governor of Pennsylvania

2. Dr. Priestley: The Philosopher and Philanthropist....

He was present and enjoyed himself, and sad must it have been to read on March 30th:

Some weeks ago, Dr. Priestley having caught cold by attending a meeting of the Philosophical Society on a wet evening, was taken ill of a violent inflammatory complaint which rendered his recovery for a long time dubi [Pg 145] ous. We announce with sincere pleasure the returning health of a man, whose life hath hitherto been sedulously and successfully devoted to the interests of mankind.

He had, indeed, been very ill. The trouble was pleurisy. Dr. Rush was his physician. By his order the patient was bled profusely seven times. During this trying and doubtful period there came to him a cheery letter from President Jefferson who had only learned of his illness. Among other things the President wrote—

Yours is one of the few lives precious to mankind, and for the continuance of which every thinking man is solicitous. Bigots may be an exception.... But I have got into a long disquisition on politics when I only meant to express my sympathy in the state of your health, and to tender you all the affections of public and private hospitality. I should be very happy to see you here (Washington). I leave this about the 30th to return about the 25th of April. If you do not leave Philadelphia before that, a little excursion hither would help your health. I should be much gratified with the possession of a guest I so much [Pg 146] esteem, and should claim a right to lodge you, should you make such an excursion.

But Priestley journeyed homeward on April 13th, and en route wrote the following letter, addressed to John Vaughan, Esq. 179 Walnut Street, Philadelphia, Pa.:

April 17, 1801
Reading, Friday Evening

Dear Sir,

I have the pleasure to inform you, agreeably to your kind request, that we are safely arrived at this place, my daughter better than when we left Philadelphia, and as to myself, I feel just as well, and as able to bear any fatigue, as before my late illness. This, however, will always remind me of your friendly attentions, and those of your sister, if a thousand and other circumstances did not do the same, and of them all I hope I shall ever retain a grateful remembrance.

Along the whole road I am struck with the marks of an astonishing degree of improvement since I came this way four years ago. I do not think that any part of England is better cultivated, and at present the wheat is in a very promising state. [Pg 147] I wish we may hear of that of England promising as well. Three years of such a scarcity is more than any country could bear, and you will believe me when I say that, if it was in my power, I would guard it not only from famine, but from every other calamity.

With my daughter's kindest remembrance, I am, as ever

Dear Sir

Yours sincerely,
J. PRIESTLEY. [8]

Resuming his correspondence with his numerous friends in England, he said:

My chief resource is my daily occupation.

He also wrote Dr. Rush his thanks for having advised him to read Noah Webster's *Pestilential Disorders* which follow the appearance of meteors and earthquakes, taking occasion also to excuse his opposition to blood-letting,—

[Pg 148]

I believe that I owe my life to your judicious direction of it. I shall never forget your so readily forgiving my suspicion, and my requesting the concurrence of Dr. Wistar after the third bleeding. It was his opinion as well as yours and Dr. Caldwell's, that my disorder required several more; and the completeness of my cure, and the speediness of my recovery, prove that you were right. In the future I shall never be afraid of the lancet when so judiciously directed.

To Rush he confided his doubts about his paper on Dreams. He cannot account for them, hence he has offered merely an hypothesis, and continues—

I frequently think with much pleasure and regret on the many happy hours I spent in your company, and wish we were not at so great distance. Such society would be the value of life to me. But I must acquiesce in what a wise providence has appointed.

His friends continued sending him books. And how joyously he received them. At times he would mention special works, as for example,—

Please to add Gate's Answer to Wall, and Wall's Reply; Sir John Pringle's Discourses [Pg 149] and Life by Dr. Kippis; Chandler's Life of King David; Colin Milne's Botanical Dictionary, Botanic Dialogues, and other books of Natural History; Kirwan's Analysis of Mineral Waters; Crosby's History of English Baptists.

In one of his letters he observed—

A person must be in my situation ... to judge of my feelings when I receive new books.

Strangely enough a *box* of books was sent him to Carlisle (Pa.) and had been there for two years before he learned of it.

Perhaps a word more may be allowed in regard to the paper on *Pestilential Disorders* by Noah Webster. This was the lexicographer. Priestley thought the work curious and important, but the philosophy in it wild and absurd in the extreme. And of Rush he asks —

Pray is he (Webster) a believer in revelation or not? I find several atheists catch at everything favourable to the doctrine of *equivocal generation*; but it must be reprobated by all who are not.

[Pg 150]

Chemists will be glad to hear that

The annual expense of my laboratory will hardly exceed 50 pounds, and I think I may have done more in proportion to my expenses than any other man. What I have done here, and with little expense, will in time be thought very considerable; but on account of the almost universal reception of the new theory, what I do is not, at present, attended to; but Mr. Watt and Mr. Kier, as good chemists as any in Europe, approve of my tract on *Phlogiston*, and truth will in time prevail over any error.

And to another he said,

Having had great success in my experiments in this country ... I shall never desert philosophy.

The following year (1802) had several points of interest in connection with the good Doctor; for one, who has followed his career thus far, will wish to call him that.

Communications from the home country and from France, while not so numerous, were yet full of interesting news. His friend Belsham [Pg 151] brought out his Elements of Philosophy of the Mind, and although Priestley paid it a most gracious tribute he did not hesitate to suggest alterations and additions of various kinds. His dearest friend Lindsey fell seriously ill this year. This gave him inexpressible anxiety and grief. As soon as Lindsey was, in a measure, restored the fraternal correspondence was resumed.

Much time was given by the Doctor to reading and preparing for the press the volumes of his *Church History* and *Notes on the Scrip-*

tures. The printing was to be done in Northumberland. Some doubt was entertained as to whether he would have funds sufficient to pay for the publication, and when the urgent letters from friends tempted him to undertake a European trip he generally replied that he was too far advanced in life, that the general debility produced by pernicious ague rendered him unfit for extended travel, and then he offset the disappointment by saying that the expense of the voyage would more than suffice for the printing of one of his proposed four volumes of the *Church History*. This was a most complete, interesting and instructive work. Even today one profits by its perusal and an immense fund of worthwhile information and knowledge may be derived from even a cursory study of his *Notes on the Scriptures*. [Pg 152]

The monotony of village life was broken by occasional letters from President Jefferson. These were most affectionate and also illuminating on national matters. Copies of these were sent to English friends with the injunction not to show them or permit them to fall into other hands.

Dr. Thomas Cooper was not with Priestley in this year (1802), being detained at Lancaster where the Assembly sat. Naturally Cooper made himself conspicuous, and Priestley prophesied a great future for him, providing that the jealousy entertained for foreigners did not prove too serious an obstacle.

Priestley took much pleasure at this period in his garden, and wrote,

Plants, as well as other objects, engage more of my attention than they ever did before.... I wish I knew a little more botany; but old, as I am, I learn something new continually.

Now and then he mentions a considerable degree of deafness, and sent to Philadelphia for a speaking trumpet, but cheerily adds,

I am, however, thankful that my eyes do not fail me.

[Pg 153]

Here and there occur plaints like these:

Though my philosophical labours are nearly over, I am glad to hear what is passing in that region in which I once moved, though what I then did seems for the present to be overlooked and forgotten. I am confident, however, as much as I can be of anything, that notwithstanding the almost universal reception of the new theory, which is the cause of it, it is purely chimerical, and cannot keep its ground after a sufficient scrutiny, which may be deferred, but which must take place in time. I am glad to find that Mr. Cruikshank in England, as well as chemists in France, begin to attend to my objections, though the principal of them have been published many years; but, as you say, many will not read, and therefore they cannot know anything that makes against the opinions they have once adopted. Bigotry is not confined to theology.

The experimental work for the year was not very great. Probably this was the result of his general physical weakness and in part it was due to his preoccupation with literary labours. How [Pg 154] ever, he did write out his results, obtained on heating "finery cinders and charcoal" and thus emphasized the gaseous product of which he observes —

It cannot be denied, however, that this gaseous oxyd of carbon (CO) is *inflammable* ... and is essentially different from all other oxyds, none of which are combustible.

Along in the month of November he wrote a vigorous protest against Cruikshank's explanation of the mode of formation of carbon monoxide. In this polemic he of course threw into prominence his precious phlogiston, the presence of which seemed unnecessary — but this was not so thought by the Doctor, who also favored the *Medical Repository* with observations on the conversion of iron into steel, in which there is but a single reference to phlogiston, but unfortunately this single reference spoils the general argument and the correct and evident interpretation of the reaction. It reads as follows:

Iron is convertible into steel by imbibing only *phlogiston* from the charcoal with which it is cemented.

[Pg 155]

There are abundant correct observations. Their interpretation sadly enough is very false, all because of the persistent introduction of phlogiston where it was not essential.

Priestley advised Rush that because of an unhealthy season he had suffered very much from ague, and said,—

Tho' I was never robust, I hardly knew what sickness was before my seizure in Philadelphia, but the old building has since that had so many shocks, that I am apprehensive it will ere long give way. But I have abundant reason to be satisfied, and shall retire from life *conviva satur*.

Devotion to work was on the part of Priestley, something marvelous. As his son and daughter-in-law were drawn to Philadelphia in February, 1803, they carried their father with them. He was rather indisposed to this, yet he disliked remaining alone at home notwithstanding the printing of the Church History required considerable personal attention. The marvelous part of it all was that while in Philadelphia, on this his fourth and last visit, while he fraternized with congenial souls and even presented himself at various social functions, he yet found leisure to print his little volume entitled "Socrates and Jesus Compared," [Pg 156] which gave much pleasure to President Jefferson, so much indeed that he hoped Priestley would,—

take up the subject on a more extended scale, and show that Jesus was truly the most innocent, most benevolent, the most eloquent and sublime character that has ever been exhibited to man.

Jefferson's genuine approval of his effort was balm to Priestley's soul. He, of course, wrote Lindsey and Belsham about it; yes, copied the letter of Jefferson and sent the same to them with the comment,—

He is generally considered as an unbeliever. If so, however, he cannot be far from us, and I hope in the way to be not only *almost*, but *altogether* what we are.

It was February 28, 1803, that the august members of the American Philosophical Society resolved:

That this Society will dine together on Saturday next, and that J. B. Smith, Wistar, Williams, Hewson & Vaughan be a Committee to make the necessary arrangements for that purpose and to request Dr. Priestley's company, informing him that [Pg 157] the Society are induced to make the request from their high respect for his Philosophical Labours & discoveries, & to enjoy the more particular pleasure of a social meeting—The Dinner to be prepared at the City Tavern or Farmer's Hotel.

It was this resolution which caused notices, such as the following to go out to the distinguished membership of the venerable Society—

Philadelphia, March 2, 1803

Sir: You are hereby invited to join the other members of the American Philosophical Society, in giving a testimony of respect, to their venerable associate Dr. Joseph Priestley, who dines with them on Saturday next at Francis' Hotel—Dinner on table at 3 o'clock.

C. Wistar
J. Williams
J. R. Smith
T. T. Hewson
J. Vaughan
Committee

An answer will be called for tomorrow morning.

DR. RUSH

[Pg 158]

It was a very dignified and brilliant company. Law, medicine, theology, science, commerce represented by very worthy and excellent gentlemen. And, among them sat the modest, unassuming, versatile Priestley. That he was happy in his surroundings there is ample reason to believe. He loved to be among men. He, too, was appreciated and eagerly sought because of his winning ways, his tolerance and liberality. He was moderately convivial though

He said that one glass of wine at dinner was enough for an old man, but he did not prescribe his own practice as an universal rule.

About eight weeks were spent in the City. On return to the dear country home the doctor took up his various duties and burdens, but the infirmities of age were often alluded to by him, and they no doubt delayed all of his work, which was further aggravated by a dangerous fall on his left hip and strain of the muscles of the thigh.

He was extremely lame and for some time went about on crutches, which held him out of his laboratory. To him this was very trying. But he persisted. He was truly a splendid example for the younger aspirants for scientific honors. During the year [Pg 159] he entered on a controversial article with his old friend Erasmus Darwin upon the subject of *spontaneous combustion*, and subsequently communicated to the *Medical Repository* an account of the conversion of salt into nitre. He had positive knowledge of this fact for quite a little while, and upon the occasion of a visit by Dr. Wistar, told the latter concerning this with the request that no mention be made of it, evidently that he might have opportunity for additional confirmation. However, very unexpectedly, Dr. Mitchill published something of a similar character, therefore Priestley believing that he ought "to acquaint experimentalists in general with all that I know of the matter," announced that in 1799 when experimenting on the formation of air from water,

having made use of the same salt, mixed with snow, in every experiment, always evaporating the mixture the salt was recovered dry. I collected the salt when I had done with it, and put it into a glass bottle, with a label expressing what it was, and what use had been made of it.

Subsequently he treated this salt, after many applications of it, with sulphuric acid, when he remarked [Pg 160] —

I was soon surprized to observe that *red vapours* rose from it.

An examination of another portion of the salt showed —

that when it was thrown upon hot coals ... it burned exactly like nitre.

So it was a conversion of sodium chloride into sodium nitrate. That this change must have come from the *snow* with which it had been dissolved, could not be doubted, and he further observed —

Now in the upper regions of the atmosphere ... there may be a redundancy of inflammable air ... and a proportion of dephlogisticated air. In that region there are many electrical appearances, as the *aurora borealis*, falling stars &c; in the lower parts of it thunder and lightening, and by these means the two kinds of air may be decomposed, and a highly dephlogisticated nitrous acid, as mine always was, produced. This being formed, will of course, attach itself to any *snow* or *hail* that may be forming ... confirm [Pg 161] ing in this unexpected manner, the vulgar opinion of nitre being contained in snow.

This seems to be the last communication of this character which came from the Doctor's pen.

He was in despair relative to the academy which had ever been his hope for the College which in his early years in Northumberland he prayed might arise and in which he would be at liberty to particularly impart his Unitarian doctrines.

An interesting item relative to the Academy appeared in the *Aurora* for April 1st, 1803. It shows that State aid for education was sought in those early days. It is a report, and reads —

A REPORT of the Committee to whom was referred the Petition of Thomas Cooper, on behalf of the Northumberland Academy, praying legislative aid. The report states that Thomas Cooper appeared before the Committee and stated that upward of $4000 had been expended on the building appropriated to that institution. That the debts due thereon amounted in the whole to near $2000. That Dr. Joseph Priestley had the power of disposing of a very valuable library consisting of near [Pg 162] 4000 volumes of scarce and well chosen books in various branches of literature and science, to any public seminary of learning in the United States, which library, the said Dr. Priestley was desirous of procuring as a gift to the Northumberland Academy, provided that institution was likely to receive substantial assistance from the legislature, so as to be enabled to fulfil the purposes of its establishment,

That the Trustees would have no occasion to ask of the legislature on behalf of that Academy, a subscription greater than a few individuals had expended, and were still ready and desirous of contributing thereto; and suggest it to your Committee, that if out of the monies due from the County of Northumberland to the State a sufficient sum was granted to exonerate the Academy from debt, no more would be wanted in the future to effect the purposes of that institution, than a sum equal in amount to the value of the library proposed to be furnished by Dr. Priestley; such value to be fixed by a person appointed for the purpose by the legislature.

The Committee was of the opinion that it would be expedient for the legislature to [Pg 163] coincide with the suggestion of Thomas Cooper and so recommended to the Legislature. Their report was adopted, 39 to 31. It was strongly advocated by Jesse Moore, Esq., General Mitchell and N. Ferguson from the city. It was opposed by Jacob Alter from Cumberland, who declared that although there were a great many public schools and colleges and places of that kind scattered over the State, he never knew any good they did, except to breed up a set of idle and odious lawyers to plague the people!

At this particular time there still existed confiscated land from the sale of which revenue was derived, and this income it had been agreed upon should be devoted to the erection and support of academies throughout the State. Later this scheme was discontinued. But, Dr. Priestley was not so enthusiastic as formerly. He was occupied with the Church History, three volumes of which were in print, and it was expected that the fourth volume would follow shortly thereafter. However, his health was precarious. He could not eat meats, and lived chiefly on broths and soups, saying, [Pg 164] —

The defect is in the stomach and liver, and of no common kind. If I hold out till I have finished what I have now on hand, I shall retire from the scene, satisfied and thankful.

This was written in August, and the Doctor stuck bravely to his literary labors. A few months later he wrote Lindsey, —

I really do not expect to survive you.

Yet, he also entertained the thought that he might, —

assist in the publication of a whole Bible, from the several translations of particular books, smoothing and correcting them where I can.

January of 1804 brought him many interesting, splendid and valuable books from friends in London. He was overjoyed on their arrival. Promptly he gave himself to their perusal because his deafness confined him to home and his extreme weakness forbade any excursions. Then the winter kept him from his laboratory, and his sole occupation was reading and writing. He entertained a variety of plans, proceeding with [Pg 165] some but in the midst of these tasks of love—in the very act of correcting proof, he quietly breathed his last! It was Monday, February 6, 1804, that Thomas Cooper, the devoted friend of Priestley, wrote Benjamin Rush:—

Dear Sir:

Mr. Joseph Priestley is not at present in spirits to write to his friends, and it falls to my lot therefore to acquaint you that Dr. Priestley died this morning about 11 o'clock without the slightest degree of apparent pain. He had for some time previous foreseen his dissolution, but he kept up to the last his habitual composure, cheerfulness and kindness. He would have been 71 the 24th of next month. For about a fortnight there were symptoms of dropsy owing to general debility: about two days before his death, these symptoms disappeared, and a troublesome cough came on perhaps from a translation to the chest.

Yesterday he had strength enough to look over a revise of the *Annotations* he was publishing on the Old and New Testament, and this morning he dictated in good language some notices which he wished his son Mr. Priestley to add to his unpublished [Pg 166] works. I am sure you will sincerely regret the decease of a man so highly eminent and useful in the literary and philosophical world, and so much presumably your friend.

Yes, the valiant old champion of a lost cause was no more. Two days before his death "he went to his laboratory"—but, finding his weakness too great, with difficulty returned to his room. Loyal to his science to the very end!

To American chemists he appeals strongly because of his persistent efforts in research. His coming to this country aroused a real interest in the science which has not waned in the slightest since his demise.

When the sad news reached the Hall of the American Philosophical Society, Dr. Benjamin Smith Barton was chosen to eulogize Priestley. This notable event took place on January 3rd, 1805. The *Aurora* reported:

Dr. Benjamin Smith Barton, one of the vice-presidents of the American Philosophical Society, having been previously appointed by the society to deliver an eulogium to the memory of their late associate, Dr. Joseph Priestley, the same was accordingly de [Pg 167] livered in the First Presbyterian Church in this city, on Thursday the 3rd inst. before the society, who went in a body from their hall to the church, preceded by their patron, the governor of the state. Invitations were given on this occasion to the Revd. Clergy of the city; the college of Physicians; the Medical Society; the gentlemen of the Bar, with the students at Law; the trustees and faculty of the University of Pennsylvania, with their students in the Arts and in Medicine; the judges and officers of the federal and state Courts; the foreign ministers and other public characters then in the city; the mayor; aldermen and city councils: the trustees and session of the First Presbyterian Church; the directors of the City Library; the directors and Physicians of the Pennsylvania Hospital, of the Alms House, and of the Dispensary; the proprietor and Director of the Philadelphia Museum; and the contributors towards the Cabinet and Library of the Society. After the conclusion of a very interesting eulogium, the society returned their thanks to the orator, and requested a copy for the purpose of publication.

[Pg 168]

One's curiosity is quickened on thinking what Barton said in his address. Search in many directions failed to bring forth the Eulogium. It had been ordered to be printed in the Transactions of the Society. This was never done. But there was a minute (seven years later) in the meeting of the Society (Nov. 6, 1812) to the effect that

Dr. Barton's request for permission to withdraw it (Eulogium) to be enlarged and published separately was referred for consideration to the next meeting.

The request was granted at the next meeting, but nowhere among Barton's literary remains was the precious document to be found. Lost very probably — when it might have revealed so much.

Priestley's death was deeply mourned throughout the land. The public prints brought full and elaborate accounts of his life, and touching allusions to the fullness of his brilliant career. Such expressions as these were heard, —

As a metaphysician he stands foremost among those who have attempted the investigation of its abstruse controversies.

As a politician he assiduously and successfully [Pg 169] laboured to extend and illustrate those general principles of civil liberty which are happily the foundation of the Constitution of his adopted Country,—

His profound attention to the belles-lettres, and to the other departments of general literature, has been successfully exemplified among his other writings, by his lectures on oratory and criticism, and on general history and policy,—

Of the most important and fashionable study of *Pneumatic Chemistry* he may fairly be said to be the father.

He was a man of restless activity, but he uniformly directed that activity to what seemed to him the public good, seeking neither emolument nor honour from men. Dr. Priestley was possessed of great ardour and vivacity of intellect.... His integrity was unimpeachable; and even malice itself could not fix a stain on his private character.

And what a splendid tribute is contained in the following passages from Cuvier: [Pg 170]

Priestley, loaded with glory, was modest enough to be astonished at his good fortune, and at the multitude of beautiful facts, which nature seemed to have revealed to him alone. He forgot that her favours were not gratuitous, and if she had so well explained herself, it was because he had known how to oblige her to do so by his indefatigable perseverance in questioning her, and by the thousand ingenious means he had taken to snatch her answers from her.

Others carefully hide that which they owe to chance; Priestley seemed to wish to ascribe all his merit to fortuitous circumstances, remarking, with unexampled candour, how many times he had profited by them, without knowing it, how many times he was in possession of new substances without having perceived them; and he never dissimulated the erroneous views which sometimes directed his efforts, and from which he was only undeceived by experience. These confessions did honour to his modesty, without disarming jealousy. Those to whom their own ways and methods had never discovered anything called him a simple worker of experiments, without method and without an [Pg 171] object "it is not astonishing," they added, "that among so many trials and combinations, he should find some that were fortunate." But real natural philosophers were not duped by these selfish criticisms.

Many encomiums like the preceding—yes, a thousandfold—could easily be gathered if necessary to show the regard and confidence held for this remarkable man to whom America is truly very deeply indebted.

Some years ago the writer paid a visit to the God's Acre of Northumberland. He arrived after dark and was conveyed to the sacred place in an automobile. Soon the car stopped. Its headlights illuminated the upright flat stone which marked the last resting place of the great chemist, and in that light not only was the name of the sleeper clearly read but the less distinct but legible epitaph:

Return unto thy rest, O my soul, for the Lord hath dealt bountifully with thee. I will lay me down in peace and sleep till I wake in the morning of the resurrection.

Pondering on these lines there slowly returned to mind the words of Franklin's epitaph, [Pg 172] —Franklin, who, years before, had encouraged and aided the noble exile, who was ever mindful of the former's goodness to him:

The Body
of
Benjamin Franklin
Printer
(Like the cover of an old book
Its contents torn out
And stript of its lettering and gilding)
Lies here food for Worms
But the work shall not be lost
For it will (as he believed) appear
once more
In a new and more elegant Edition
Revised and corrected
by

The Author

And then, by some strange mental reaction, there floated before the writer the paragraph uttered by Professor Huxley, when in 1874 a statue to Priestley was unveiled in the City of Birmingham:

Our purpose is to do honour ... to Priestley the peerless defender of national [Pg 173] freedom in thought and in action; to Priestley the philosophical thinker; to that Priestley who held a foremost place among the 'swift runners who hand over the lamp of life,' and transmit from one generation to another the fire kindled, in the childhood of the world, at the Promethean altar of science.

FOOTNOTES:

[1] Chemistry in Old Philadelphia, J. B. Lippincott Co., Phila., Pa.

[2] Correspondence of Priestley by H. C. Bolton, New York, 1892.

[3] Mr. Berthollet discovered that oxygenated muriatic gas, received in a ley of caustic potash, forms a chrystallizable neutral salt, which detonates more strongly than nitre.

[4] Nine Famous Birmingham Men—Cornish Brothers, Publishers, 1909.

[5] James Woodhouse—A Pioneer in Chemistry—J. C. Winston Co., Phila.—1918.

[6] James Woodhouse—A pioneer in Chemistry—J. C. Winston Co., Phila.—1918.

[7] See *Chemistry in America*, Appleton & Co. and *Chemistry in Old Philadelphia*, The J. B. Lippincott Co., Philadelphia, Pa.

[8] The original of this letter is now the property of Dr. C. A. Browne, New York. He graciously permitted it to be inserted here.

www.ingramcontent.com/pod-product-compliance
Lightning Source LLC
Chambersburg PA
CBHW031447210526
45464CB00005B/2360